Disclaimer

The publisher of this book is by no way associated with the National Institute of Standards and Technology (NIST). The NIST did not publish this book. It was published by 50 page publications under the public domain license.

50 Page Publications.

Book Title: Model Development and Validation for Particle Release Experiments in a Two-story Office Building

Book Author: William S. Dols; Andrew K. Persily; Jayne B. Morrow;

Book Abstract: Whole-building airflow and contaminant transport modeling has a potentially important role in the development of contaminant sampling strategies in response to the airborne release of chemical or biological agents. The effectiveness of these strategies relies on the ability of the selected sampling locations to adequately characterize the levels of contamination throughout an exposed facility to a desired level of confidence in the sampled results. The Department of Homeland Security has sponsored a series of multi-agency exercises, during which contamination experiments were performed to gauge the confidence that could be obtained by various sampling strategies as well as the effectiveness of various sampling methods in a real-world setting. These experiments are very resource intensive and time-consuming, limiting the number of experiments that can be reasonably performed. Building simulation can be used to perform virtual experiments that would allow more tests to be performed under a much larger set of building operational and environmental configurations. However, in order for the simulations to be useful, the building models need to provide realistic results with a high level of confidence. The purpose of this report is to describe a simulation validation effort based on measurements of contaminant levels performed during the aforementioned exercises.

Citation: NIST TN - 1717

Keywords: aerosol; agent; biological; modeling; multizone modeling; particle; response; validation

NIST Technical Note 1703

Model Development and Validation for Particle Release Experiments in a Two-story Office Building

W. Stuart Dols
Andrew K. Persily
Jayne B. Morrow

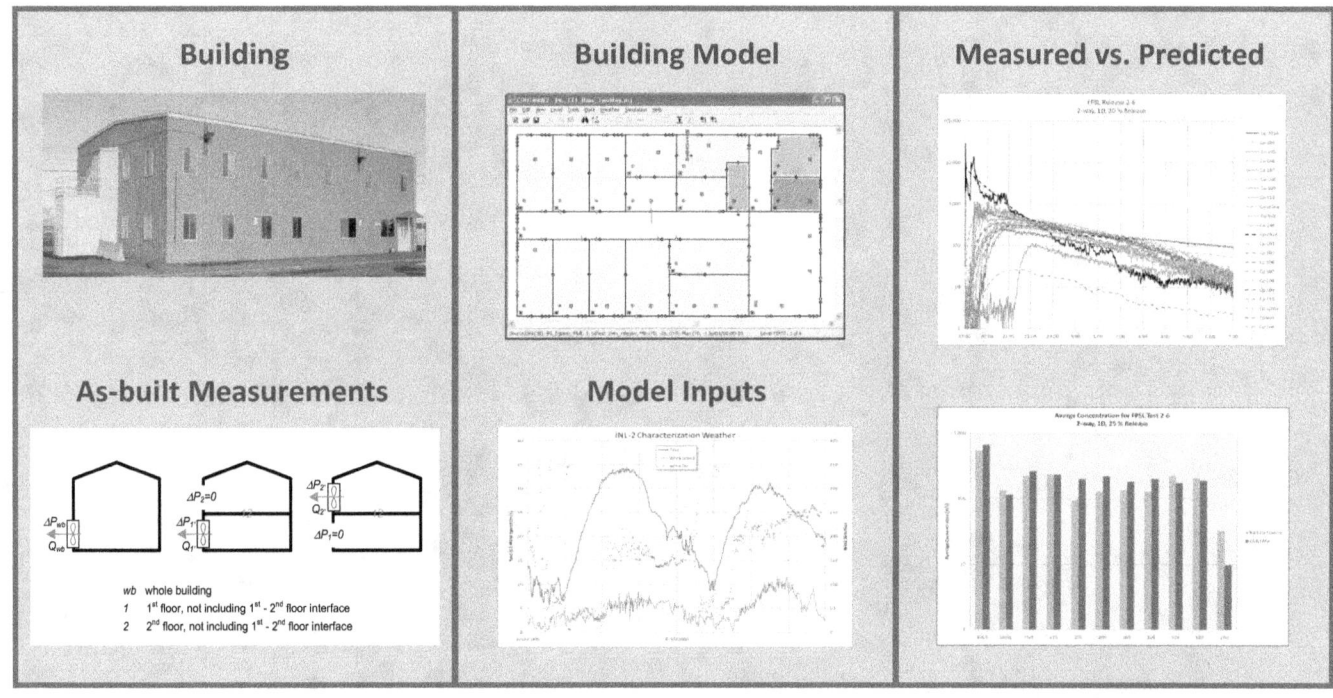

NIST National Institute of Standards and Technology • U.S. Department of Commerce

NIST Technical Note 1703

Model Development and Validation for Particle Release Experiments in a Two-story Office Building

W. Stuart Dols
Andrew K. Persily
Jayne B. Morrow

October 2011

U.S. Department of Commerce
Rebecca Blank, Secretary (Acting)

National Institute of Standards and Technology
Patrick D. Gallagher, Director

Certain commercial entities, equipment, or materials may be identified in this document in order to describe an experimental procedure or concept adequately. Such identification is not intended to imply recommendation or endorsement by the National Institute of Standards and Technology, nor is it intended to imply that the entities, materials, or equipment are necessarily the best available for the purpose.

National Institute of Standards and Technology Technical Note 1703
Natl. Inst. Stand. Technol. Tech. Note 1703, 97 pages (October 2011)
CODEN: NSPUE2

Abstract

Whole-building airflow and contaminant transport modeling has a potentially important role in the development of contaminant sampling strategies in response to the airborne release of chemical or biological agents. The effectiveness of these strategies relies on the ability of the selected sampling locations to adequately characterize the levels of contamination throughout an exposed facility to a desired level of confidence in the sampled results. The Department of Homeland Security has sponsored a series of multi-agency exercises, during which contamination experiments were performed to gauge the confidence that could be obtained by various sampling strategies as well as the effectiveness of various sampling methods in a real-world setting. These experiments are very resource intensive and time-consuming, limiting the number of experiments that can be reasonably performed. Building simulation can be used to perform virtual experiments that would allow more tests to be performed under a much larger set of building operational and environmental configurations. However, in order for the simulations to be useful, the building models need to provide realistic results with a high level of confidence. The purpose of this report is to describe a simulation validation effort based on measurements of contaminant levels performed during the aforementioned exercises.

Two sets of experiments were performed in a two-story office building at the Idaho National Laboratory (INL) with the goal of exercising sampling methods and strategies and determining the level of confidence in the results. Measurements performed during these experiments were also used to validate a whole-building airflow and contaminant transport model of the building, the results of which are presented herein. Although the experiments were not carried out with the intent of providing model validation data, they still resulted in a rich set of particle measurement data to be used for this purpose.

The multizone building airflow and contaminant transport modeling software, CONTAM, developed by the National Institute of Standards and Technology (NIST), was used in this validation effort. A CONTAM model of the INL facility was initially developed based on design documentation and then refined based on measurements of the as-built building including building pressurization tests and ventilation system airflow rate measurements. This report provides a detailed description of the process by which the building modeling assumptions were refined based on comparisons with two detailed sets of experimental results, and then how the refined model was applied to the remaining two dozen experimental test cases. Results were evaluated based on *ASTM D5157 Standard Guide for Statistical Evaluation of Indoor Air Quality Models*.

Generally, the results were not all within recommended levels presented in the ASTM guide, but they were encouraging in light of the fact that several aspects of the building itself were not well-characterized for input to the building model, e.g., wind effects associated with a large containment tent surrounding the building. Qualitatively, the results yielded similar contamination patterns within an order-of-magnitude of measured results. The study highlights the importance of properly characterizing model inputs in order to improve confidence in simulation results.

Key Words: aerosol; agent; biological; modeling; multizone modeling; particle; response; validation

Table of Contents

1 Introduction

The Department of Homeland Security (DHS) Science and Technology Directorate is tasked with establishing the scientific, engineering, and technological resources to enhance the security of the United States, while leveraging existing resources to support technological advancement with standards for all chemical, biological, radiological, nuclear and explosive threats. Detection technology performance specifications and test and evaluation standards are critical to establishing a standards infrastructure to provide confidence in detection technologies. However, of equal importance is to provide confidence in our ability to find and recover contamination through validated sample collection and handling protocols. Methods used by the Centers for Disease Control and Prevention (CDC) and the Agency for Toxic Substances and Disease Registry (ATSDR) to determine the extent of *B. anthracis* spore contamination after the 2001 attacks demonstrated a range in recovery efficiencies for the various methods utilized including HEPA vacuum socks, wipe materials and swabs (dry and wet) for spore recovery from nonporous surfaces [1]. In 2005 the Government Accountability Office (GAO-05-251) published a report calling for an increase in confidence in the ability of federal agencies to detect potential biological contamination events as well as the performance of sampling strategies and uncertainties associated with sampling methodologies. More recently in 2008, the GAO reported (GAO-08-180) that the response community's ability to predict and model airborne dispersion of chemical, biological, radiological and nuclear (CBRN) materials as well as define the area of contamination with rapid detection technology capabilities to be placed in the hands of first responders was lacking. In response to a 2005 report DHS organized the Validated Sampling Plan Working Group (VSPWG), composed of multiple federal agencies and national laboratories to ensure that the overall process of sampling activities has been validated. Since then the procedures recommended by the CDC to collect spores from nonporous surface using swabs or wipes and wetting agents (solutions used to moisten the wipe materials) have been through multi-lab validation by the response laboratories[1, 2] and the agencies have worked closely to develop guidance and sample collection strategies. However, as pointed out by the GAO, modeling capabilities and statistical analysis can be used to provide sampling strategists with guidance for sample site selection and estimate recovery efficiencies. For example, modeling can increase our understanding of the role of building characteristics on contaminant distributions throughout the building and effective sampling strategies.

During the years 2007 and 2008, a facility at Idaho National Laboratory (INL) served as the location for a multi-agency effort to experimentally evaluate the performance of sampling strategies to characterize and clear a building after a bio-contamination event [3, 4]. Both time and cost are significant constraints in large-scale field tests such as those conducted at INL. Modeling provides a means to expand our understanding of the distribution and transport of biological agents in a building beyond what can be explored experimentally given the physical and resource limitations of experimental testing. Model parameters can be adjusted to study the impact of variations in physical parameters that cannot be addressed experimentally in either a practical or economic manner.

In a previous study, the CONTAM multizone building airflow and contaminant transport modeling tool [5] was integrated with the sample planning tool VSP [6] to demonstrate the ability to perform relatively low-cost virtual experiments and evaluate sample selection strategies [7]. While these virtual experiments can provide an economical accompaniment to actual experiments, there are valid concerns as to the validity of their application. Therefore, this project was undertaken to develop a model of the INL facility and perform simulations of some of the cases that were run at the facility. It is important to note that while a rich set of experimental data was collected during these tests, the experiments were not performed with the intent of providing model validation data such as temperature, pressure, air change rate, wind pressure effects of the surrounding decontamination tent, etc. Nevertheless, this exercise was valuable in relation to the methods by which one would develop a model of a building that has undergone a contamination event for which there is incomplete characterization of model inputs.

In this work, the National Institute of Standards and Technology (NIST) developed a model of the INL test building (PBF-632) based on as-built building properties. This model was adapted from one that was developed as part of the previous study [7] and based on available design documentation and various engineering assumptions. NIST conducted an on-site inspection and performed measurements of required model inputs including building envelope leakage, inter-floor leakage, and ventilation airflows. The model was then verified against measured leakage data and tuned to achieve a more-reliable level of performance. Data from the INL tests and the building model with as-built parameters were then used to simulate contaminant release scenarios in an attempt to validate the CONTAM model for biological contaminant dispersal modeling capability.

The end users of validated sampling methodologies and particulate modeling capabilities are the first responder community and the government entities that characterize and clear contaminated facilities post-aerosol release. Guidelines, standards and materials are utilized by first responder professionals throughout the various phases of response including state and local HazMat crews, National Guard Bureau Civil Support Teams, United States Coast Guard or other military personnel as well as responders from the federal agencies including the CDC National Institute for Occupational Safety and Health, the United States Environmental Protection Agency, and the Federal Bureau of Investigation. The quality of sample collection strategies and sample plan development will increase with validation of model performance and of sample collection plans.

2 Building Description

In order to develop a CONTAM model of the building, information on the building layout, geometry and ventilation system are needed. Design information was obtained from mechanical system drawings and electronic floor plans (i.e., DWG AutoCAD files). A limited set of on-site measurements was also performed to obtain as-built information to define the building model within CONTAM.

2.1 Design

The building used for this study is PBF-632 located at INL and shown in Figure 1. This is the same building used in exercises aimed at the evaluation of sample planning methods [3, 4]. Floor plans for the two floors of PBF-632 are shown in Figure 2 and Figure 3. Each floor is approximately 24.4 m x 15.2 m for a total of 372 m^2 per floor. As designed, each floor contains a constant volume air handler located within a mechanical room on the floor it serves. Outdoor air is brought in through an intake duct on the return air side of each air handler. Supply air ducts are located above suspended ceilings on the floor which they serve, and the space above the suspended ceilings also serves as a return air plenum. Return ducts draw air out of these plenums from just above each of the two mechanical equipment rooms. Supply air is provided to all occupied spaces except the restrooms and janitorial closet. Two return air grilles are located in the ceilings along the main hallway of each floor. Dedicated exhaust fans serve the four restrooms (two on each floor) and each of the mechanical rooms. Total design supply airflows for the 1st and 2nd floors are 1,200 L/s and 1,180 L/s respectively. Design restroom exhaust flows are 94 L/s for each restroom and 47 L/s for each mechanical room. The design outdoor air intake rates are unavailable.

Figure 1 – INL Building PBF-632

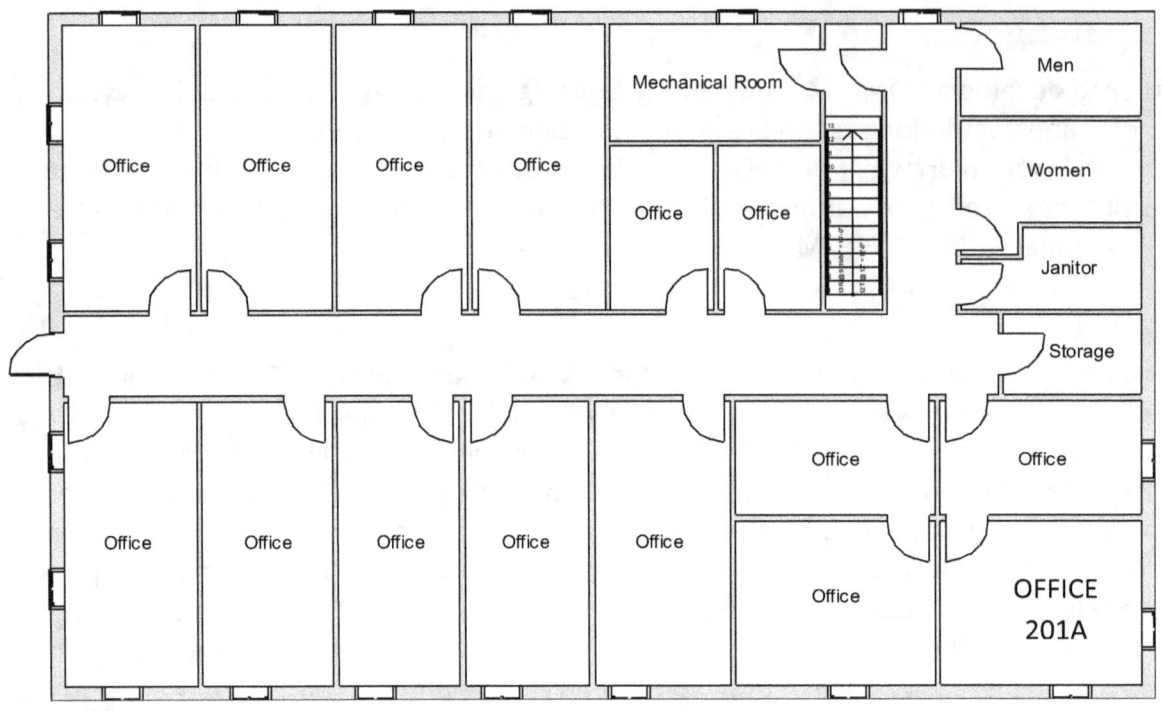

Figure 2 – 2nd Floor Plan

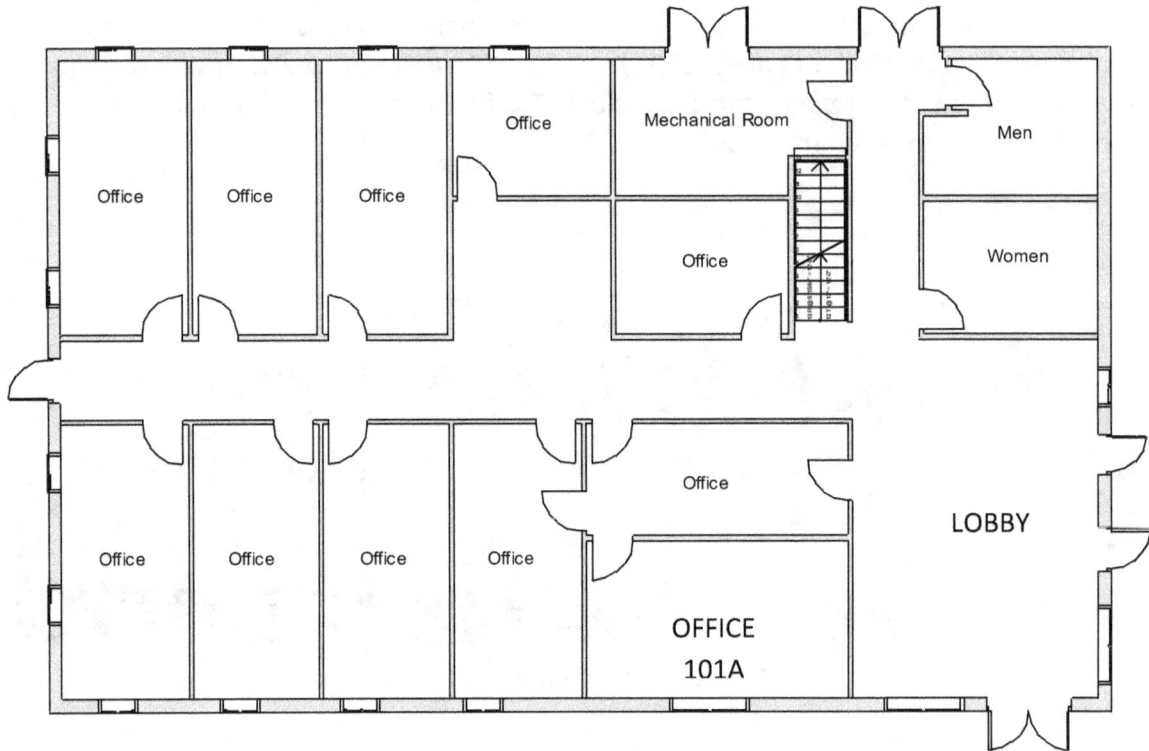

Figure 3 – 1st Floor Plan

4

2.2 Building Modifications

The building had been "decommissioned" prior to the study and was significantly modified in order to perform the experiments to demonstrate and validate response and recovery sampling approaches and technologies [3, 4]. A large tent was erected around the entire building in order to contain the chemicals used to decontaminate the building between test events (see Figure 4). Rigid fiberglass insulation was applied to all suspended ceilings and the edges were sealed with metal duct tape (see Figure 5). This sealing was done in order to minimize air and contaminant transport between the first and second floors, so that each floor could be utilized for individual experiments without contaminating the other. Dedicated return ducts and decontamination distribution ducts were installed below the suspended ceiling, i.e., within the occupiable space, on each floor (see Figure 5). Return air ducts were fitted with removable bag filters (MERV 11) at the inlets to the ducts. Reversible flow fans were installed within the return air ducts. Remote controllers for these reversible fans and the original air handlers were installed within a trailer located adjacent to the building. Outdoor air intake ducts were capped off inside the mechanical rooms, so that no outdoor air was introduced directly into the ventilation system. The decontamination system was located adjacent to the building but within the containment tent.

Figure 4 – Decontamination containment tent surrounds the building

| (a) | (b) |

Figure 5 – (a) Rigid, foil-faced insulation attached to ceilings and (b) Retrofitted decontamination and return air ducts

3 Measurements

Two series of tests were performed at the INL facility to investigate issues related to sampling methods and strategies – one during the summer of 2007 and another during the summer of 2008. The two rounds of tests are referred to as INL-1 and INL-2 respectively. These tests were carried out by a group of government agencies and contractors for the purposes of evaluating biological sampling methods and strategies [3, 4], but not for the purposes of validating airflow and contaminant transport modeling. However, a rich set of particle data was obtained during both trials.

Each round of testing included a set of tests referred to as "building characterization" experiments that were aimed at determining particle dispersion patterns for various building configuration and operation conditions, e.g., interior door position and ventilation system operation. Once these tests were completed and the desired building configuration and operating conditions were determined, the main test events were conducted using a biological simulant *Bacillus atrophaeus* (Bg).

In all, INL-1 consisted of 13 characterization tests and INL-2 consisted of 12 tests. For the purposes of this model validation study, only the building characterization tests were utilized as explained in section 3.2.2. Two of the cases from INL-2 were simulated first, cases 2-6 and 2-12. These tests included a more detailed set of measurements as explained in section 3.2. These two cases were used to "tune" the simulation parameters to improve comparisons with particle measurements. Once simulation parameters were adjusted based on these two data sets, the remaining INL-1 and INL-2 characterization cases were simulated, and the results are presented in section 5.2.3.

3.1 Building Airflow

In order to develop a CONTAM building model, building airflow characteristics are needed including ventilation system airflow rates and envelope and inter-zone leakage properties. Actual ventilation system airflow rates often differ from design values, so airflow rates were measured for the building air handling systems. Dimensions of transfer grills located in office doors were also measured, because some tests were conducted with closed doors when air would be flowing through these grilles. In order to characterize the building envelope airtightness, whole-building fan pressurization and depressurization tests were performed using a blower door according to the ASTM E779 test procedure [8]. Because some of the exterior doors would not remain closed during pressurization, only the depressurization tests were utilized to determine envelope leakage rates.

3.1.1 Ventilation system airflow rates

HVAC system supply airflow rates were measured using a hood balometer having an uncertainty of $\pm$ 34 m^3/h. The original air handler fans were turned on and hood balometers were used to measure flows from the supply air diffusers on both floors as well as the return airflows into the retrofitted return air ducts. These return air inlets were located at each end of the long rectangular duct located along the hallway of each floor (see Figure 5).

3.1.2 Envelope leakage

The building was depressurized using a blower door to determine envelope leakage rates for the building as a whole and for each floor separately. Measurements were used to determine envelope leakage areas for the 1st and 2nd floors (red and blue lines respectively in Figure 6) as well as that of the assumed pressure boundary between the two floors (intersection of red and blue boundaries). The stairway is open to the first floor, so the entire stairway was assumed to be within the first floor pressure boundary. As mentioned previously, each floor was served by a dedicated air handling/duct system that essentially "links" the occupiable space and plenums of each floor.

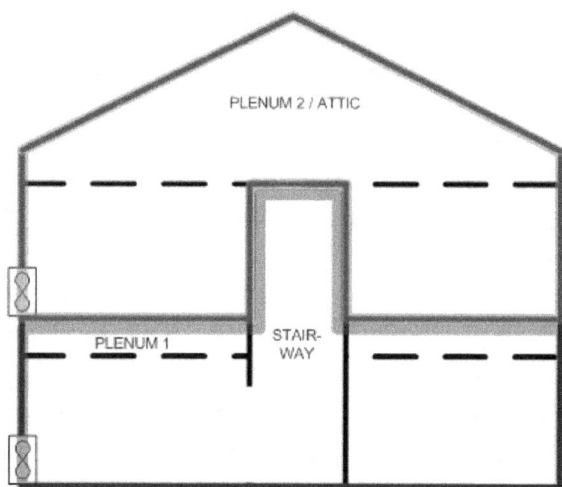

Figure 6 – Assumed pressure boundaries for blower door tests

Three tests were performed using the test configurations shown in Figure 7 in which the airflow rate Q was measured at selected pressure differences ΔP: the whole building leakage was measured with the stairway door open between the 1st and 2nd floors, the 1st floor plus the 1st-2nd interface was measured with the 2nd floor exterior door and windows open, and the 2nd floor plus the 1st-2nd interface was measured with the 1st floor exterior doors and windows open. Subscripts on each of the measurements associated with these tests are: *wb* for whole building not including the 1st-2nd interface, *1'* for 1st floor including the interface between floors, *2'* for the 2nd floor including the interface between floors.

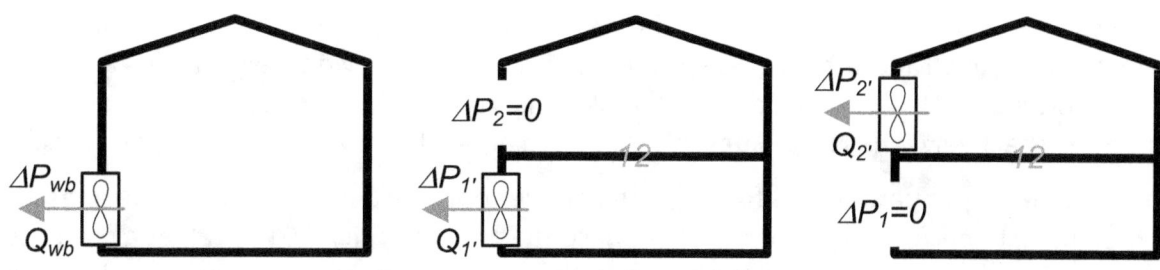

Figure 7 – Blower door test configurations

From these measured values, the leakage of the 1st-2nd floor interface, 1st floor, and 2nd floor were all estimated. Curve fits were performed on the flow versus pressure data for the three tests, fitting the data to the power law equation:

$$Q = C\,(\Delta P)^n \tag{1}$$

where Q is volumetric airflow rate through the fan, ΔP is the pressure difference across the fan, C is the air leakage coefficient and n is the pressure exponent. Results of the measurements are provided in Table 1 and have been corrected for standard conditions as provided in the ASTM E779 test procedure [8]. Table 1 provides the following values for each of the three measurements:

ELA_4 Effective leakage area at a reference pressure of 4 Pa, which is the equivalent size orifice that would provide the same airflow at a 4 Pa pressure difference based on the data fitted to equation 1.

NL_{75} Leakage airflow rate at 75 Pa normalized by above ground surface area, A_s. This value is the airflow rate that would be required to provide a 75 Pa pressure difference divided by the above ground building surface area which in this case includes four sides of the building and the roof.

NL_{50} Leakage airflow rate at 50 Pa normalized by above ground surface area. This is similar to NL_{75} but at a different reference pressure and in different units.

These values are presented for comparison to previously measured airtightness levels and recommended practice. 95 % confidence interval estimates in the ELA_4 values were determined to be approximately 4 % of the values provided in Table 1.

Measurements	ELA_4		NL_{75}	NL_{50}
	cm^2	cm^2/m^2	$L/s\text{-}m^2$	$m^3/h\text{-}m^2$
whole building (Q_{wb})	2411	2.64	3.72	10.6
1st floor ($Q_{1'}$)	2287	3.73	5.39	15.3
2nd floor ($Q_{2'}$)	1779	1.70	2.87	8.0

Table 1 – Blower door measurement results

Airtightness measurements of U.S. office buildings are presented by Emmerich and Persily [9]. According to their summary of measurements in 201 buildings performed as of 2005, the average normalized leakage rate at 75 Pa is 7.89 $L/s\text{-}m^2$ with a standard deviation of 9.94 $L/s\text{-}m^2$ and a range of 0.75 $L/s\text{-}m^2$ to 46.67 $L/s\text{-}m^2$. Leakage measurements of the INL building are well within these limits and below the average. Brennan et al. [10] point out that the British Air Tightness Testing and Measurement Association (ATTMA) [11] provide best practice and normal practice tightness levels for air-conditioned office buildings at 50 Pa to be 2 $m^3/h\text{-}m^2$ and 5 $m^3/h\text{-}m^2$ respectively. The NL_{50} values in Table 1 are above the levels provided by ATTMA.

Having obtained these parameters for the three blower door test configurations, the lines of fit were then used to perform mass balances (as described in Emmerich 2003 [12]) at a select set of points along the curves to obtain the required flows, i.e., flow across the interface between floors 1 and 2 (Q_{12}), flow across 1st floor pressure boundary excluding the 1st-2nd interface (Q_1) and flow across 2nd floor pressure boundary excluding the 1st-2nd interface (Q_2) as follows:

$$Q_{12} = \frac{(Q_{1'} + Q_{2'} - Q_{wb})}{2} \tag{2}$$

$$Q_1 = Q_{1'} - Q_{12} \tag{3}$$

$$Q_2 = Q_{2'} - Q_{12} \tag{4}$$

Results of these calculations are presented in Table 2.

Measurement	C $m^3/h\text{-}Pa^n$	n	A_S m^2	ELA_4		NL_{75} $L/s\text{-}m^2$	NL_{50} $m^3/h\text{-}m^2$
				cm^2	cm^2/m^2		
$1^{st} - 2^{nd}$ interface (Q_{12})	310	0.65	372	828	2.23	3.91	10.8
1st floor walls only (Q_1)	637	0.55	242	1462	6.05	7.72	22.3
2nd floor walls & roof (Q_2)	370	0.63	672	951	1.41	2.30	6.4

Table 2 – Leakage rates calculated from blower door measurements

Further, leakage rates for exterior doors and windows were assumed based on a library of leakage components [13] and subtracted from the values in Table 2 to obtain wall, floor and ceiling assembly leakage elements. The final values used in the as-built building models are provided in Table 6 in section 4 of this report.

3.2 Particle Measurement

Particle measurements were performed by government contractors [3, 4] using particle counters during all test cases. For a select number of cases particle deposition was also measured directly using settling vials. Specifically, settling vials were used in characterization tests 2-6 and 2-12. Sample locations for those two tests are provided in Figure 8 and Figure 9.

3.2.1 Release agents

Two release agents were used during the *building characterization* phase of INL-1: green visolite in dry powder form and fluorescent polystyrene latex (FPSL) stored in aqueous solution. Visolite powder was disseminated using a dry powder eductor system, and the FPSL was disseminated using a micro-controlled aerosol generator/nebulizer stated by the manufacturer to have a generation efficiency of 75 %. Manufacturers of both indicate that these particles were monodispersed as 1 μm particles. Only the FPSL particles were used in the INL-2 round of tests.

3.2.2 Real-time aerosol monitors

Particle sampling was performed using aerosol monitors. These monitors operated continuously to provide near real-time particle counts in a size range of 0.7 μm to 10 μm. The particle counters could be "tuned" to detect the fluorescent particles in order to eliminate interference from background particles. This ability to "tune" the counters was not implemented for the biological tracer (Bg) experiments, so those experiments were not considered for the purposes of this validation effort. Further, the particle counts obtained during the Bg tests would include particles whether they were viable or not, but sampling methods being validated, i.e., swab, wipe and vacuum, only provide counts of viable contaminants. Therefore, comparisons between the monitor counts and other sampling methods were not useful for the purposes of this model validation. Therefore, this validation study focused on the *characterization* test data using the FPSL and visolite particles, for which particle counts per liter were provided nominally every second.

Measurement locations appear as grey boxes labeled with a sample identifier in Figure 8 and Figure 9. The real-time device is proprietary and accuracy specifications for the measurements were not available [14]. This lack of uncertainty is not of great concern, because this study focused more on relative, qualitative prediction of contamination throughout the building rather than on absolute predictions. Also, the multi-zone software employed does not predict variations in contaminant concentrations within a room but rather room-averaged concentrations. Plots of the real-time measurements are provided in Figure 10 and Figure 11 for test 2-6 and 2-12 respectively. Release locations for these tests are room 101A and 201A, respectively. These two cases are provided as example data sets and will be reviewed in detail later in this report. They are plotted here to provide an idea of the nature of the real-time measurements including their fluctuations with time. These plots show data at 10 s intervals (every 10th value) as opposed to every second of data collected.

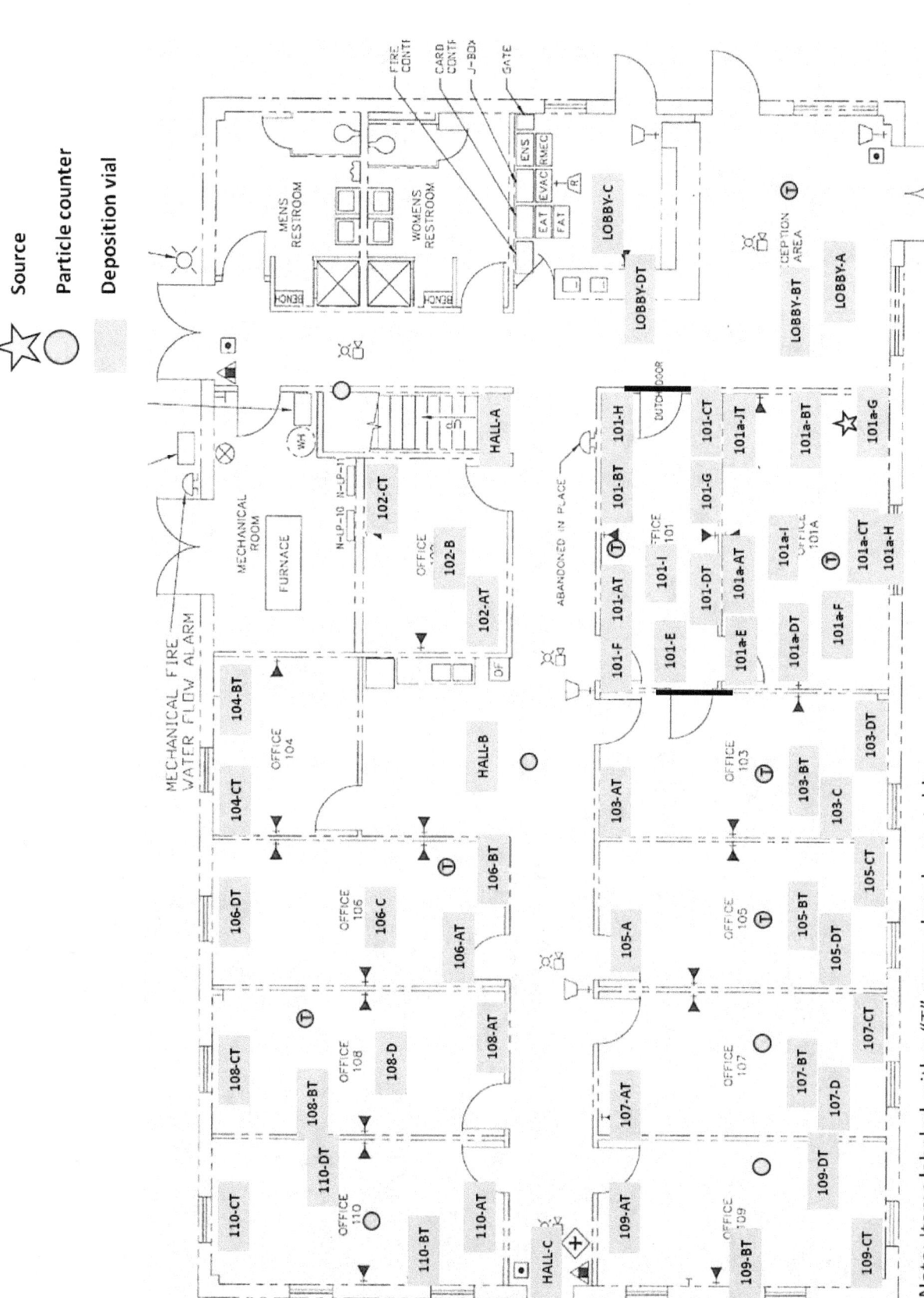

Source

Particle counter

Deposition vial

Figure 8 – First floor building plan showing sample locations for test 2-6 (provided by ICx Technologies, Inc.)

Note: Items labeled with a "T" were placed on a table

11

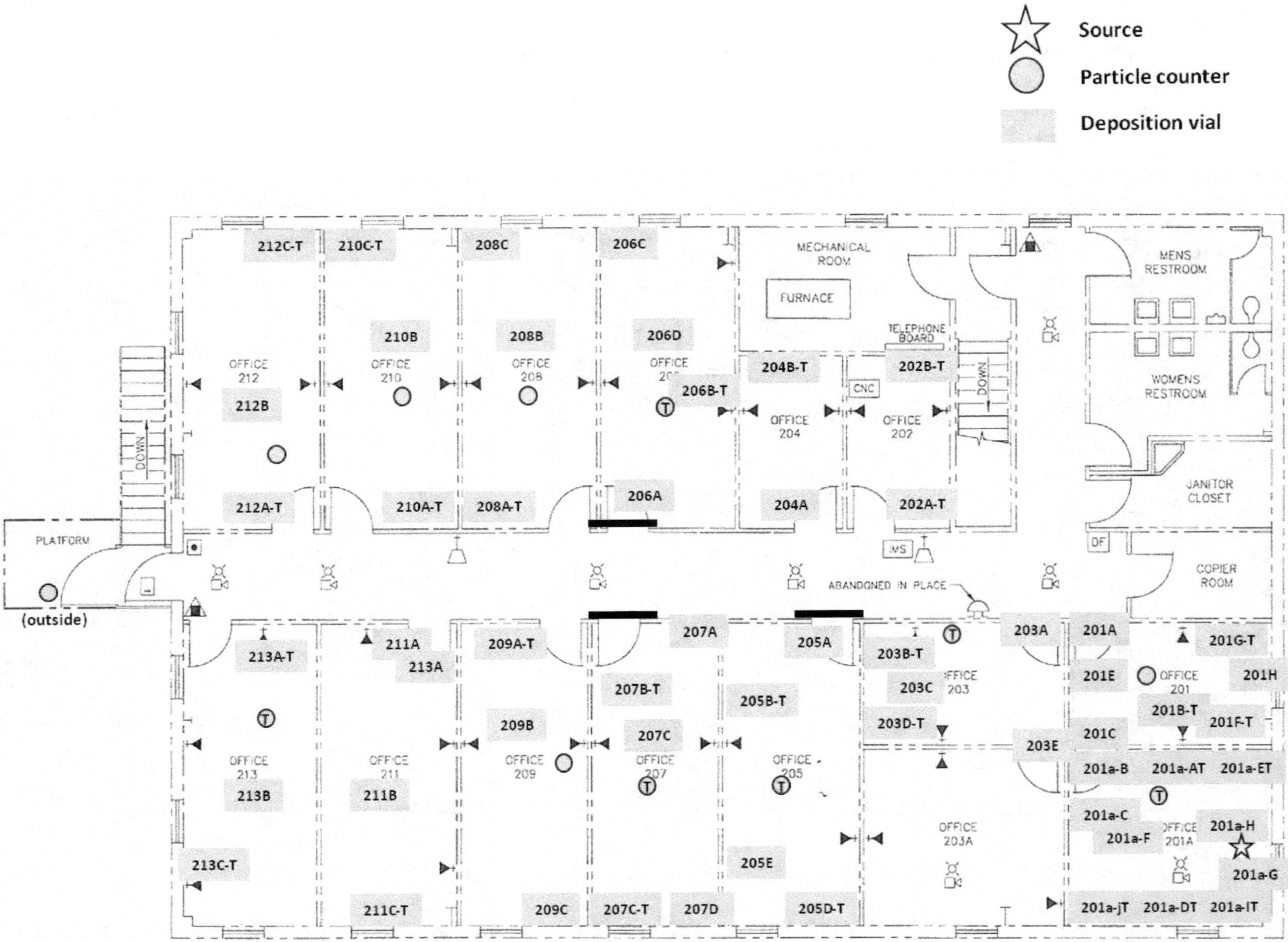

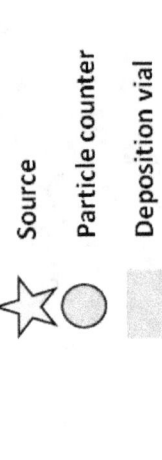

Note: Items labeled with a "T" were placed on a table

Figure 9 – Second floor building plan showing sample locations for test 2-12 (provided by ICx Technologies, Inc.)

12

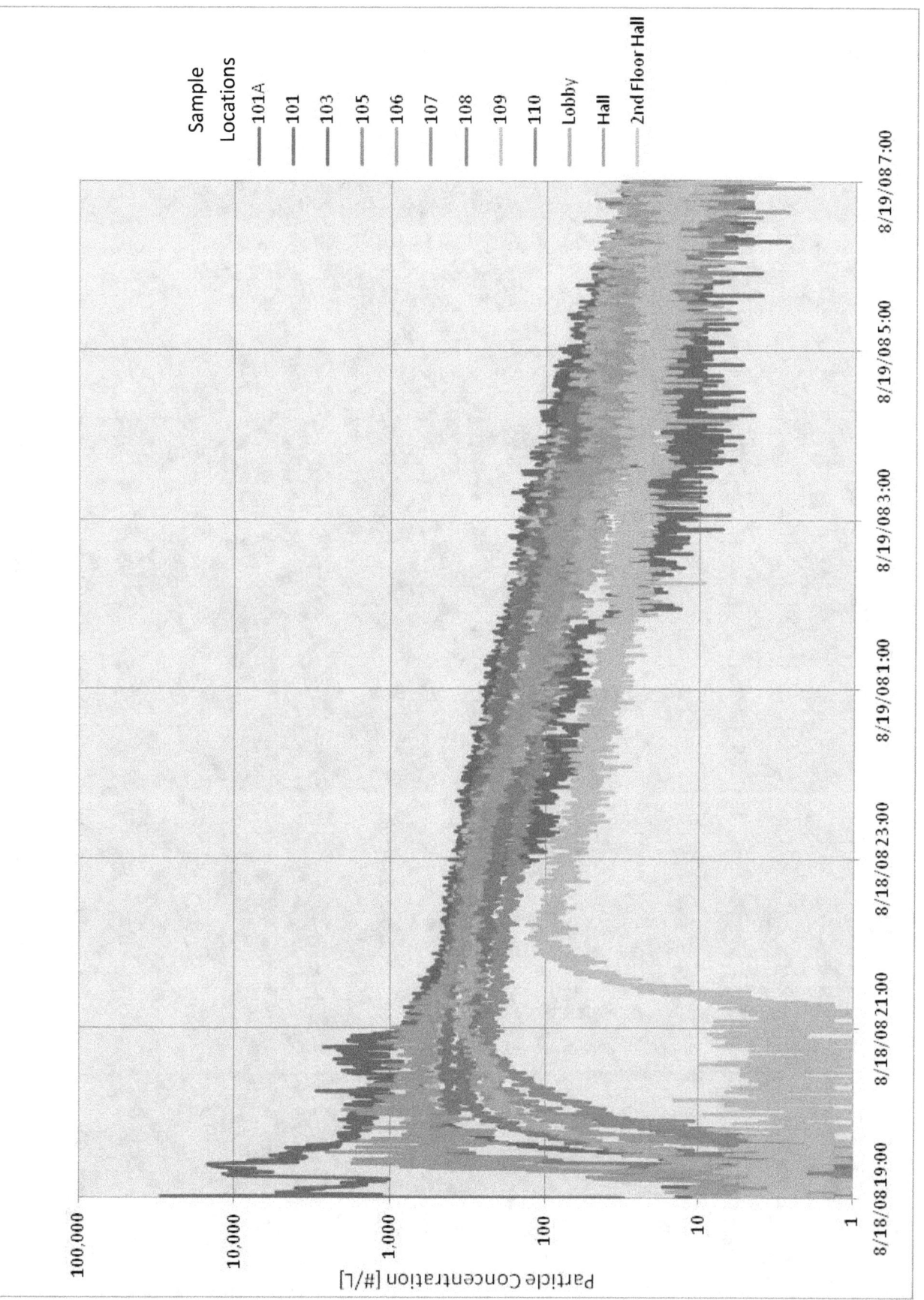

Figure 10 – Plot of real-time particle measurements for FPSL test 2-6

13

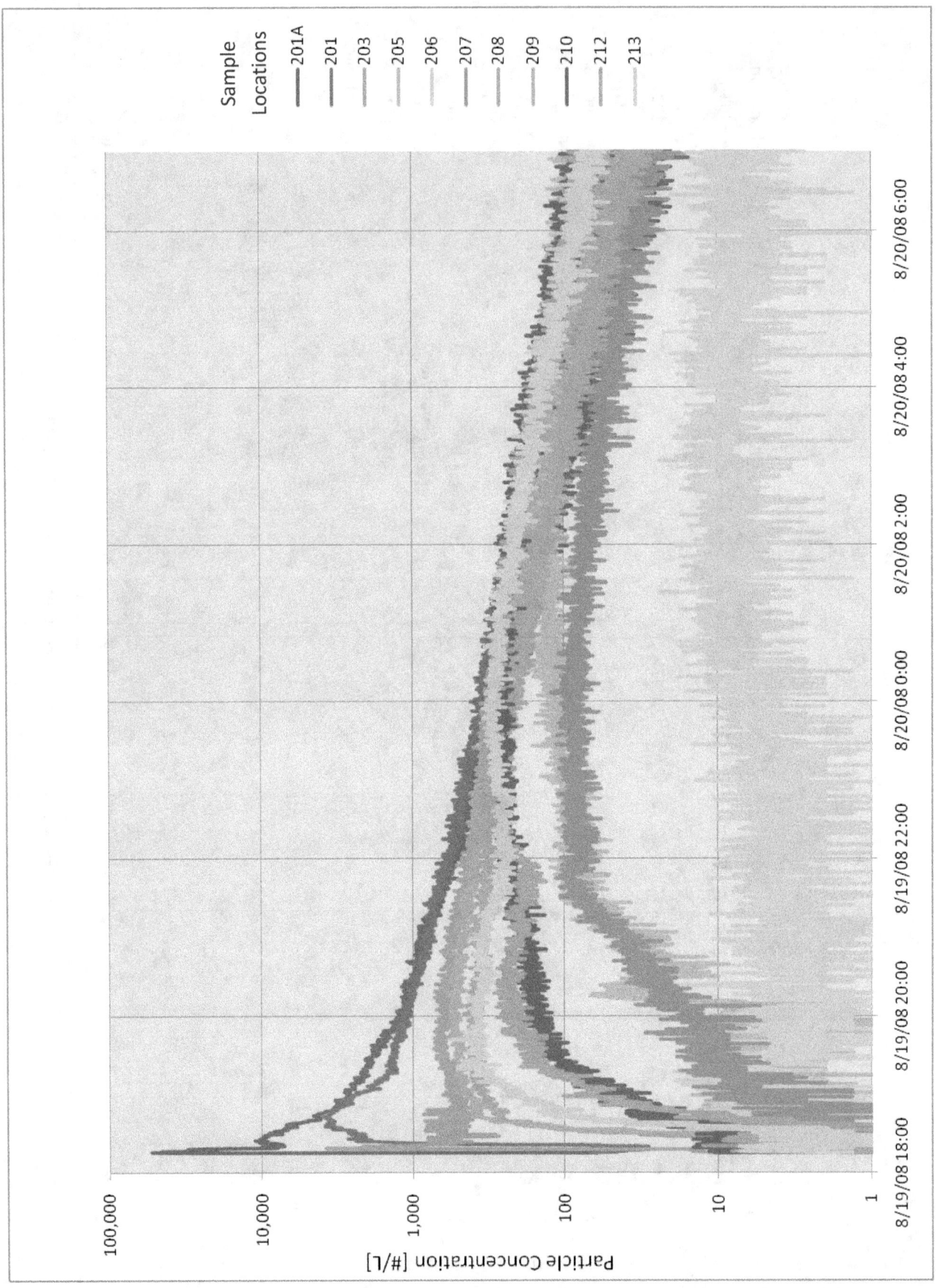

Figure 11 – Plot of real-time particle measurements for FPSL test 2-12

14

3.2.3 Deposition vials

In two cases (2-6 and 2-12) particle deposition was measured directly using small glass vials (0.78 cm^2 diameter opening) that were placed throughout the building to collect settled particles. Measurement locations appear as grey rectangles in Figure 8 and Figure 9. Particles that settled within the vials were later resuspended in aqueous solution and counted using flow cytometry with a manufacturer's stated accuracy of ± 15 %. Results are provided in Table 3 and Table 4 for each room number in which they were placed. The values for multiple samples in each room are presented in the order of their alpha-numeric identifiers along with the average, standard deviation and ratio of standard deviation to average (coefficient of variation) for each room. It is evident from these measurements that the deposition is not uniform in most cases with the coefficient of variation ranging from 50 % to 100 %.

	Vial Deposition [particles/cm^2]										
	101A	101*	102*	103	104*	105	107	108	109	Hall	Lobby
	291	87	19	106	23	47	11	31	47	175	17
	108	0	11	94	37	22	22	17	9	100	30
	537	2	39	94	61	45	11	28	31	11	12
	735	112		123		109	39	17	37		42
	16	668									
	23	223									
	3	217									
	9	69									
	6										
AVG	192	172	23	104	40	56	21	23	31	95	25
STD	273	217	14	14	19	37	13	7	16	82	14
CVAR	142 %	126 %	63 %	13 %	48 %	67 %	64 %	31 %	52 %	86 %	53 %

Table 3 – Deposition vial measurements for FPSL test 2-6 (* no real-time data)

	Vial Deposition [particles/cm^2]													
	201A	201	202*	203	204*	205	206	207	208	209	210	211*	212	213
	248	427	144	69	31	5	2	0	17	3	8	0	20	3
	53	136	217	139	36	2	3	0	20	6	3	22	0	14
	3	246		22		2	2	3	22		17	20	11	
	28	359				8	2	5						
	209	468				3		30						
	147	326												
	376	137												
	114	28												
	297													
	376													
AVG	185	266	181	77	34	4	2	8	20	5	9	14	10	9
STD	138	156	52	59	4	3	1	13	3	2	7	12	10	8
CVAR	75 %	59 %	29 %	77 %	11 %	64 %	22 %	167 %	13 %	47 %	76 %	87 %	97 %	92 %

Table 4 – Deposition vial measurements for FPSL test 2-12 (* no real-time data)

4 CONTAM Building Model

Each level of the building is represented in CONTAM by a schematic of the floor plan via the CONTAM sketchpad. The model used in this study consists of four levels – the 1st and 2nd floors and their respective plenums that contain the air distribution ductwork. CONTAM sketchpads are shown for each of the four levels in Figure 12 and Figure 13. The nominal floor area of the building is 372 m^2 with a nominal volume of approximately 2610 m^3. Individual level properties are provided in Table 5.

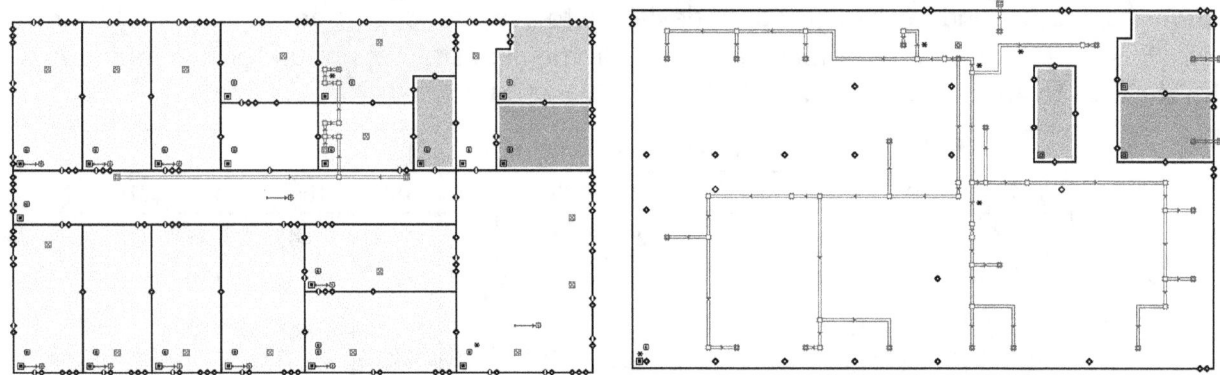

Figure 12 – CONTAM representation of 1st Floor and Plenum

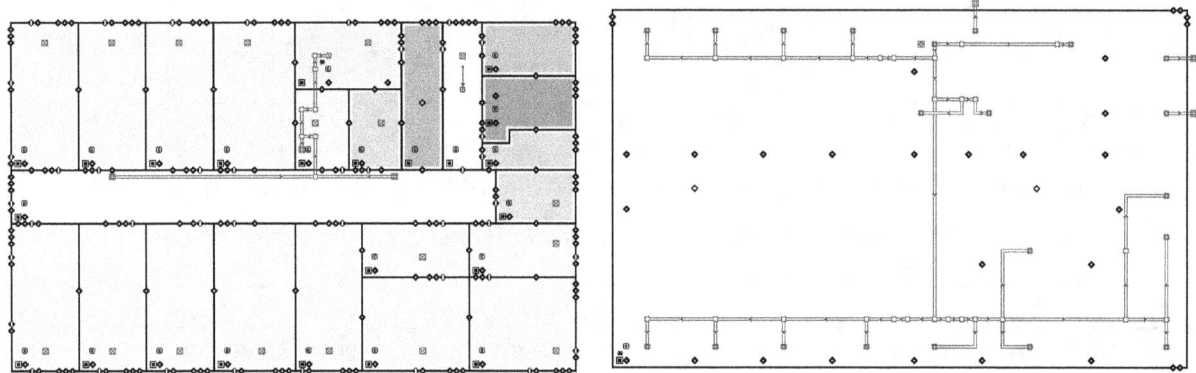

Figure 13 – CONTAM representation of 2nd Floor and Plenum

Level	Nominal Height [m]	Nominal Volume [m^3]
1st floor	2.44	906
1st floor plenum	0.61	227
2nd floor	2.44	906
2nd floor plenum/attic	1.22	570

Table 5 – Building level properties

4.1 Airflow Characteristics

4.1.1 Building component leakage

CONTAM requires information on the interconnections between zones (airflow paths) used to define the airflow model of the building. These are provided in the form of CONTAM *airflow elements* that define the mathematical relationship between pressure and airflow for a given building component. Exterior wall components, ceiling and floor leakages were determined based on the blower door tests described previously. Other elements were selected from a library of leakage components [13]. These airflow properties are provided in Table 6 in the form of leakage areas at a given reference pressure as described in the ASHRAE Handbook of Fundamentals [15]. Unless otherwise stated, the leakage components had a reference pressure of 4 Pa, discharge coefficient of 1.0 and flow exponent (n) of 0.65.

Building Component	Leakage Properties
exterior wall – 1st floor	5.59 cm^2/m^2 n = 0.60
exterior wall – 2nd floor	4.38 cm^2/m^2 n = 0.60
window frame – horizontal slider	0.80 cm^2/m
exterior door frame – double door	8.00 cm^2/m^2
exterior door frame – single door	1.00 cm^2/m^2
metal roof system	0.0019 cm^2/m^2 n = 0.5
interior wall	2.65 cm^2/m^2
interior door undercut	95.25 cm^2/m
interior door louver	1363 cm^2
interior door – open (Two-way flow model)	2.032 m x 0.8128 m, C_D = 0.78, $\Delta T_{two-way}$ = 0.01 °C
ceiling	0.4 cm^2/m^2
floor	1.75 cm^2/m^2

Table 6 – Airflow properties of the CONTAM building model

Blower door tests provide the combined leakage of various "bulk" sections of the building, e.g., exterior walls of the 1st floor. This combined leakage is then distributed horizontally across the entire section of the building in order to provide interconnectivity between the outdoor and indoor spaces. Further, the exterior wall leakage is distributed vertically in order to account for buoyancy flow (i.e., stack effect). Similarly, the inter-floor leakage is apportioned throughout the floors/ceilings of all rooms of a floor. Where there were other obvious leaks, such as cut-outs in the floors and ceilings, specific airflow paths were provided and the combined leakages were reduced accordingly.

During the experiments, interior door opened/closed configurations were varied in an attempt to vary the levels of contamination throughout the building. Closed doors are indicated by the thick black lines across the doorways in Figure 8 and Figure 9. Two-way flow models are used to capture the inter-room mixing that occurs through large openings when temperature variations exist between rooms connected by such openings.

4.1.2 Wind effects

In order to account for wind effects on the building, a set of wind pressure coefficient profiles are implemented in the building model. Two sets of coefficient profiles are utilized – one for the longer walls of the building and one for the shorter walls. These coefficients profiles were obtained from the CONTAM library [13] and are based on correlations provided for surface-averaged wind on rectangular buildings in chapter 16 of the ASHRAE Fundamentals Handbook [15]. In order to account for the likely significant effect of the tent structure on the surface wind pressures, the building was assumed to be located in urban terrain and surface pressures adjusted accordingly. The tent provides for wind shielding that is quite atypical of that which would normally be encountered for an actual building. While it was observed that the tent is very leaky, it is likely to be a fairly significant source of uncertainty with respect to modeling wind effects on the building pressures.

Wind speed and direction data were downloaded from the "PBF" weather station of the NOAA INL Weather Center website [http://www.noaa.inel.gov/metgraph] for the days for which simulations were performed (plots provided in Appendix D). Because the building heating and air-conditioning system was not operating (with the exception of the air handlers), the effects of buoyancy due to inside-outside temperature difference were omitted from the simulation by setting both temperatures to be constant at 20 °C. Therefore, the predicted infiltration rates are purely a function of wind-driven airflow. Indoor temperatures were not measured during any of the bio-release experiments.

4.1.3 Mechanical ventilation system

The full duct system of the building was modeled in order to better account for ventilation system airflows while the system fans were off, which was the case for many tests. Duct dimensions were obtained from design documents and measurements made in the building. The duct system was balanced, using CONTAM's duct balancing feature, based on airflow measurements performed in the building and presented previously.

4.2 Release Agent

The release agent was modeled as a particle having an equivalent diameter of 1.0 μm. Particle deposition surfaces were included in each zone and assumed to consist of only the zone's floor area. A constant deposition velocity of 3.50 x 10^{-5} m/s was assumed for all deposition surfaces within the CONTAM model. This deposition velocity is based on the settling velocity of a 1 μm particle [16]. In "typical" indoor, mechanically ventilated spaces, deposition onto horizontal surfaces dominates for particles of this size and the deposition rate is largely unaffected by the friction velocity and hence turbulence intensity according to the model presented by Lai and Nazaroff [17].

5 Building Simulations

A number of simulations were performed with the first being performed to verify the modeled envelope leakage characteristics. Once the leakage characteristics were established to be in reasonable agreement with the measured values, the contaminant release simulations were performed. Only those tests that utilized the non-biological agents were simulated as explained earlier. Initially, two contaminant releases from INL-2 were simulated, cases 2-6 and 2-12. These two cases were first simulated using the *well-mixed* assumption of CONTAM. Comparisons were made between the measured and predicted results using methods presented in section 5.2.1. Based on these comparisons, simulation parameters and assumptions were adjusted to better account for contaminant transport, e.g. flow through open doors and transport delays in hallways and ducts, which improved agreement between measured and predicted results relative to the *well-mixed* modeling assumption. The remaining cases were then simulated using the adjusted assumptions and parameters and the results are presented in section 5.2.3. This section describes the process of refining the modeling assumptions and the differences in results that are obtained based on the different sets of modeling assumptions.

5.1 Building Pressurization Tests

Simulations were first performed to verify the modeled airflow characteristics of the building. This was done by using the model to simulate the blower door tests that were performed and then comparing the simulated results to the measured results. Simulation results matched the measured blower door data to within ± 6 % over a range of 0 Pa to 25 Pa. Exact agreement was not obtained due in part to the adjustment of the "general" wall leakage rates to account for the intentional openings of which component leakage measurements were not performed. The leakage of the intentional openings, i.e., windows and exterior doors, were instead based on a library of leakage components [13].

5.2 Contaminant Release Tests

The first two cases simulated were INL-2 characterization tests 2-6 and 2-12 for which deposition was measured with the deposition vials. Source locations were rooms 101A and 201A with release amounts of 1 mg and 2 mg respectively of FPSL. Releases were very short in duration, so they were modeled as CONTAM burst sources, which occur essentially instantaneously. Releases occurred at 7:00 PM and 6:15 PM on consecutive days for test 2-6 and 2-12 respectively. In both cases, the particles were allowed to settle overnight, and the ventilation systems were not turned on during the tests.

5.2.1 *Methods of comparison between measurements and simulations*

Comparisons were made between measured and simulated results using both qualitative and quantitative methods. Qualitative methods included visual inspection of plots containing both real-time particle concentrations and simulated results. In order to quantify the agreement between the measured and simulated results, comparisons were made based on *ASTM D5157 Standard Guide for Statistical Evaluation of Indoor Air Quality Models* [18] (referred to herein as the ASTM guide). In addition, other comparisons were made between the simulations and measurements including contaminant peak values and timing.

Contaminant concentrations

The ASTM guide provides six quantitative indicators to be used together to evaluate model performance. The guide also notes that plots of predicted and measured concentrations over time as well as the plotting of residuals can be used for qualitative evaluation. The quantitative parameters and their recommended levels to provide adequate model performance are provided in Table 7.

Parameter	Range
Correlation coefficient, r	$r \geq 0.9$
Regression slope, m	$0.75 \leq m \leq 1.25$
Regression intercept, b	$b/\overline{Co} \leq 0.25$
Normalized mean square error, $NMSE$	$NMSE \leq 0.25$
Fractional bias, FB	$|FB| \leq 0.25$
Fractional bias of variance, FS	$|FS| \leq 0.50$

Table 7 – ASTM Guide D5157 model evaluation parameters

The first three parameters are regression indicators that relate to the *goodness of fit* of a line to the plot of Cp versus Co. A line with a slope of 1.0, intercept of 0.0 and a correlation coefficient of 1.0 would indicate perfect agreement between the two sets of data. The following equations for the last three parameters are provided from the ASTM guide, wherein the terms Co and Cp represent observed (measured) concentration and predicted concentration respectively and σ^2 the variance:

$$NMSE = \overline{(Cp - Co)^2} / (\overline{Cp}\ \overline{Co}) \tag{5}$$

$$FB = 2(\overline{Cp} - \overline{Co}) / (\overline{Cp} + \overline{Co}) \tag{6}$$

$$FS = 2\left(\sigma^2_{Cp} - \sigma^2_{Co}\right) / \left(\sigma^2_{Cp} + \sigma^2_{Co}\right) \tag{7}$$

Note that Fractional Bias (FB) provides a normalized range of values between ± 2.0 and is very similar to percent difference for values between ± 25 %. Values of *FB* between ± 1.636 indicate that averages are within one order of magnitude of each other, and values between ± 1.960 are within two orders of magnitude of each other.

Particle deposition

Total particle deposition was evaluated using vial deposition as the reference value. Vials were only employed for the two test cases being evaluated initially, 2-6 and 2-12. Deposition results obtained for CONTAM deposition sinks located in each room were compared directly to the vial values as were deposition amounts based on the average particle measurements in corresponding locations in which real-time aerosol monitors were located (as shown in Figure 8 and Figure 9). Average measured and simulated particle concentrations were also compared graphically. Average particles concentrations were related to deposition amounts via the previously mentioned deposition rate of 3.50 x 10⁻⁵ m/s as follows:

$$M_d = \int_{t_1}^{t_2} \rho_{air}\, v_d\, A_s\, C(t)\, dt \qquad (8)$$

where

M_d	= mass deposited [kg]
ρ_{air}	= density of air [kg/m³]
v_d	= deposition velocity [m/s]
A_s	= deposition surface area [m²]
$C(t)$	= mass fraction of contaminant at time t [kg/kg]
t	= time [s]

By the mean value theorem, $\int_{t_1}^{t_2} C(t)\, dt$ is equal to $C_{avg}\Delta t$, where C_{avg} is the average concentration over time period Δt.

5.2.2 Simulation results for cases 2-6 and 2-12 – Well-mixed assumption

Initially, simulations were performed under the well-mixed assumption and with all zones having uniform temperature throughout the building, so there were no two-way airflows between rooms with open doorways. Buoyancy-driven two-way airflows only occur with temperature differences between rooms. These simulation results are plotted with the real-time particle results in Figure 14 and Figure 15. The solid lines are the observed concentrations and the dashed lines are the predicted concentrations. The observed plots consist of 100 s running averages in order to smooth the highly-variable data that is presented in Figure 10 and Figure 11. ASTM guide statistics for these two cases are provided in Table 8 and Table 9.

The following observations are made in reviewing Figure 14 and Figure 15. In both cases, the predicted concentrations in the release location (101A and 201A) immediately after the contaminant releases are greater than the measured values. The measured values tend to drop relatively quickly and remain well below the predicted values, but both values then decay at similar rates for some time thereafter. The cause of the discrepancy between the measured and predicted initial concentration is not clear. Uncertainty in the aerosolization efficiency of the particle generator is one explanation. As noted earlier, the manufacturer's claimed efficiency was 75 %, but that value was not verified as part of these tests. Also, the particle counters measured concentration at a single point in the room, which may not be representative of the room average concentration. There are also significant delays in the time at which elevated contaminant levels occur for the predicted values (relative to observed values) in several of the rooms, e.g., rooms 109 and 209. Peak magnitude and timing were evaluated and are presented in Figure 16 and Figure 17. The bars on these plots provide the peak timing in the order that the observed peaks appeared. It is evident that many of the predicted peaks occur at significantly different times than the observed peaks in both tests. Also, the magnitudes of the predicted peaks are often not even within an order of magnitude of the observed values, but perhaps more importantly, they did not exhibit the same order of occurrence.

The bold values in Table 8 and Table 9 indicate those results that are within the recommended specifications of the ASTM guide. Very few indicators are within the acceptable limits. Only with one exception, values of the regression intercept (b) divided by the average observed concentration ($\overline{Co}$) are within recommended limits. However, as is discussed in the ASTM guide, it is very important to consider all the criteria together as each is meant to address different statistical characteristics of data correlation. For example, the locations whose values of $b/\overline{Co}$ are within recommended limits, the other regression indicators are well outside of the recommended limits including negative or near zero slopes of the regression line.

Deposition amounts and average particle concentrations, as well as FB, are compared in Table 10 and Table 11 for tests 2-6 and 2-12. Bold values are those within the levels recommended in the ASTM guide. Only a few deposition values determined from average particle counter measurements have an FB in the ± 0.25 range when compared to the average deposition values measured using the vials. Measured and predicted average concentrations also differ significantly with the absolute FB of all being well above 1.0. However, for test 2-6 most locations are within one order of magnitude ($|FB| \leq 1.64$), and in case 2-12 most are between one and two orders of magnitude with a couple greater than two orders of magnitude ($|FB| \leq 1.96$). Figure 18 and Figure 19 provide a graphic representation of the individual vial deposition measurements along with the deposition values based on the measured and predicted average particle concentrations and assumed deposition rate. For test 2-6, most of the deposition values based on the particle counter measurements fall within the range of individual vial measurements, but only one of the values based on the simulation results falls within these limits. The trend by location in average deposition vial measurements is also fairly well captured by the particle counter results. For test 2-12, none of the simulation results are within the range of vial measurements. Fewer particle counter-based deposition values are within the vial measurements as well, but the overall trend in the average vial measurements is fairly well captured.

22

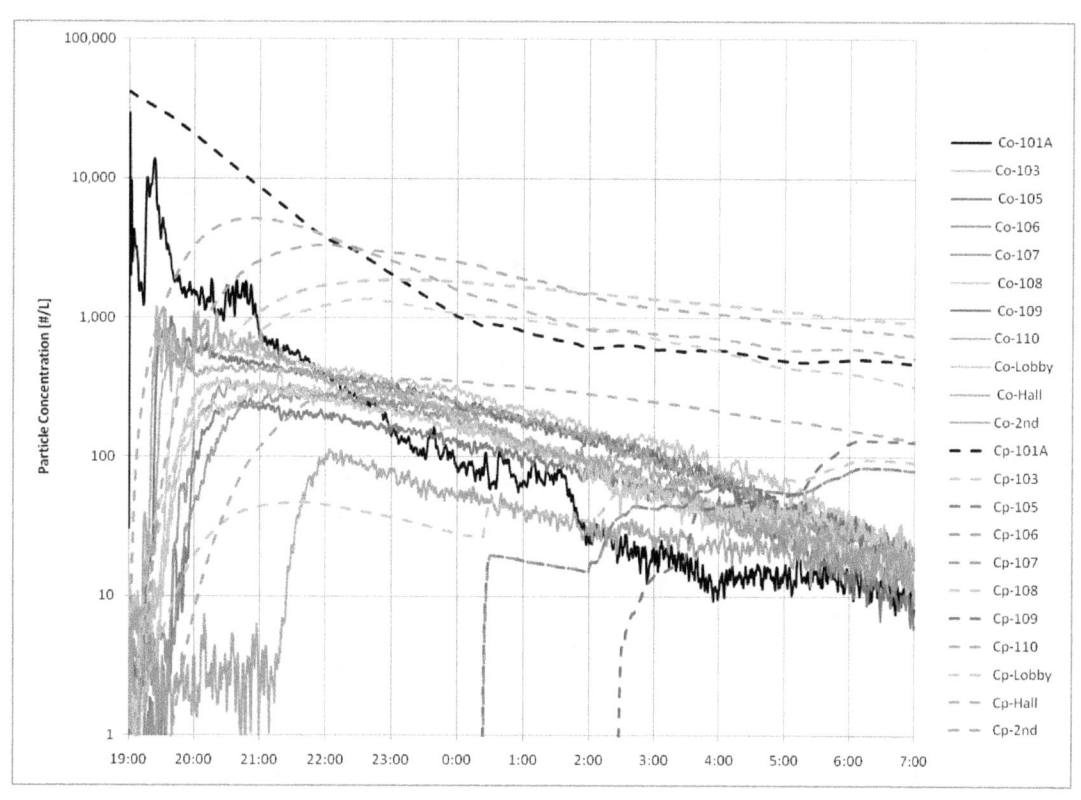

Figure 14 – Predicted (well-mixed) vs. measured particle concentrations for FPSL test 2-6

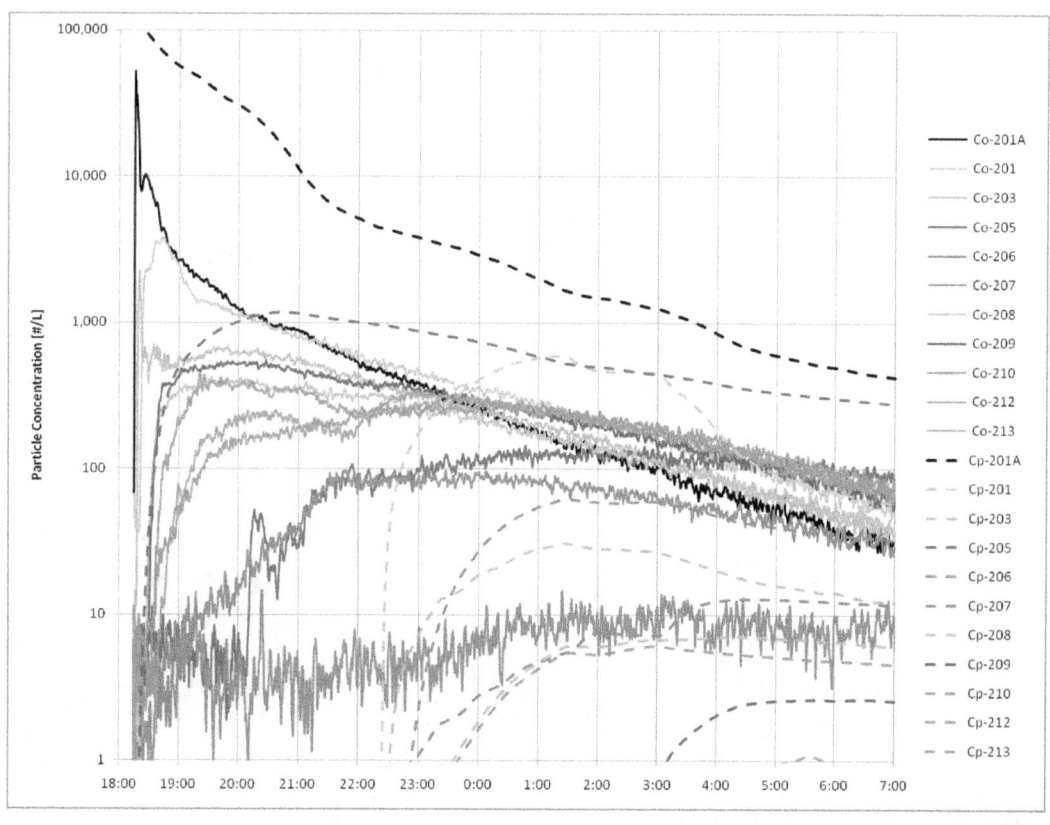

Figure 15 – Predicted (well-mixed) vs. measured particle concentrations for FPSL test 2-12

	Criteria	101A	Lobby	Hall	103	105	106	107	108	109	110	2nd
N		4321	4321	4321	4321	4321	4321	4321	4321	4321	4321	4321
r	≥ 0.9	0.74	0.70	0.75	-0.50	-0.70	0.38	-0.68	0.21	-0.56	0.58	0.76
m	0.75 - 1.25	3.48	1.94	3.75	-0.05	-0.23	1.52	-0.18	**0.86**	-0.10	2.90	2.58
$b/\overline{Co}$	≤ 0.25	4.08	2.60	2.98	**0.23**	0.47	6.72	0.34	7.34	**0.19**	3.87	2.98
$NMSE$	≤ 0.25	23.64	3.49	8.41	7.54	6.65	7.54	8.42	7.61	19.36	6.72	4.63
$|FB|$	≤ 0.25	1.53	1.28	1.48	-1.39	-1.25	1.57	-1.44	1.57	-1.67	1.49	1.39
$|FS|$	≤ 0.50	1.83	1.54	1.85	-1.95	-1.60	1.76	-1.74	1.77	-1.88	1.85	1.68

Table 8 – ASTM guide statistics for FPSL test 2-6 (well-mixed)

	Criteria	201A	201	203	205	206	207	208	209	210	212	213
N		4591	4591	4591	4591	4591	4591	4591	4591	4591	4591	4591
r	≥ 0.9	0.76	-0.42	-0.70	-0.29	0.55	-0.04	-0.85	-0.76	-0.03	-0.55	-0.70
m	0.75 - 1.25	5.33	-0.10	-0.03	-1.70	2.94	-0.01	-0.02	0.00	0.00	0.00	0.00
$b/\overline{Co}$	≤ 0.25	6.13	0.40	**0.07**	7.75	0.40	**0.09**	**0.03**	**0.01**	**0.01**	**0.00**	**0.00**
$NMSE$	≤ 0.25	41.93	8.53	43.24	5.85	3.83	14.01	94.63	528.61	84.43	1130.25	2112.28
$|FB|$	≤ 0.25	1.68	-1.09	-1.85	1.43	1.08	-1.70	-1.94	-1.99	-1.95	-2.00	-2.00
$|FS|$	≤ 0.50	1.92	-1.77	-1.99	1.89	1.87	-1.91	-2.00	-2.00	-2.00	-2.00	-2.00

Table 9 – ASTM guide statistics for FPSL test 2-12 (well-mixed)

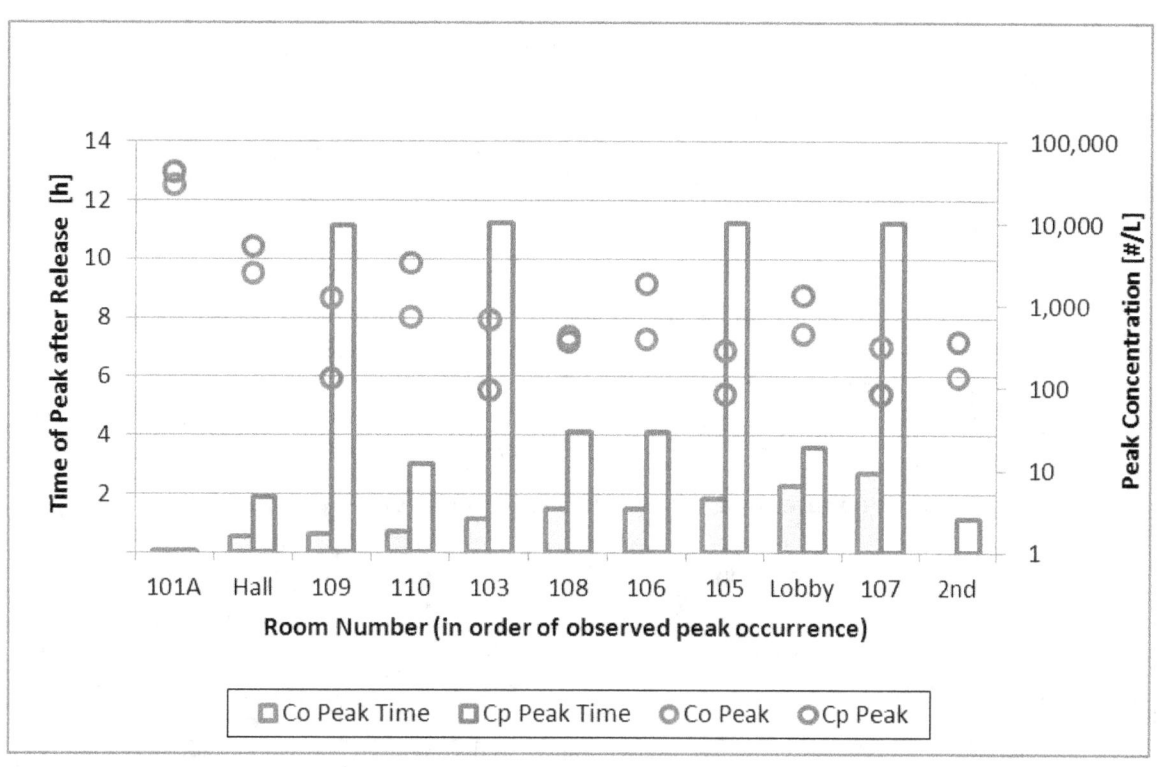

Figure 16 – Peak timing for FPSL test 2-6 (well-mixed)

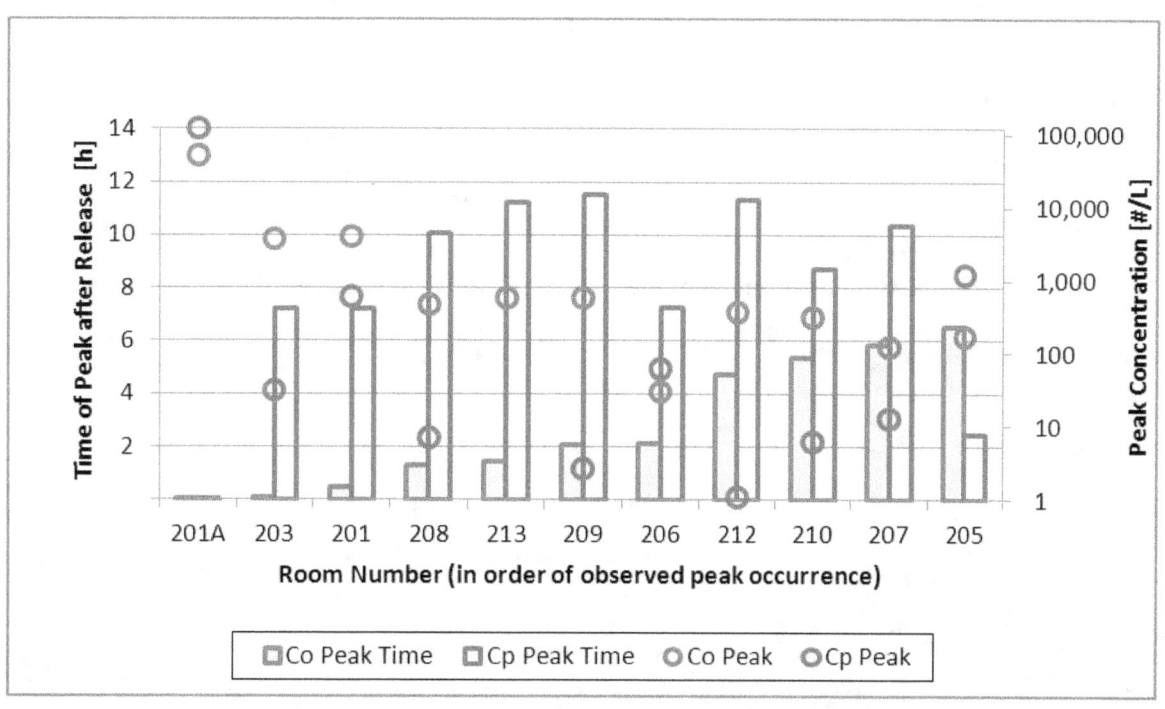

Figure 17 – Peak timing for FPSL test 2-12 (well-mixed)

Room	Deposition [#/cm^2]					Average Concentration [#/L]		
	Vial Average	Particle Counter	Fractional Bias	CONTAM	Fractional Bias	Particle Counter	CONTAM	Fractional Bias
101A	192	82	-0.80	620	1.05	543	4,102	1.53
Lobby	25	20	**-0.22**	92	1.14	134	608	1.28
Hall	95	34	-0.96	226	0.81	222	1,492	1.48
103	104	35	-1.00	6	-1.77	231	42	-1.39
105	56	14	-1.18	3	-1.77	95	22	-1.25
106	n/a	19	n/a	159	n/a	127	1,051	1.57
107	21	20	**-0.02**	3	-1.45	134	22	-1.44
108	23	19	**-0.18**	159	1.49	128	1,051	1.57
109	31	34	**0.08**	3	-1.64	223	20	-1.67
110	n/a	30	n/a	206	n/a	201	1,363	1.49
2nd	n/a	5	n/a	27	n/a	32	180	1.39

Table 10 – Deposition rates and average concentrations for FPSL test 2-6 (well-mixed)

Room	Deposition [#/cm^2]					Average Concentration [#/L]		
	Vial Average	Particle Counter	Fractional Bias	CONTAM	Fractional Bias	Particle Counter	CONTAM	Fractional Bias
201A	185	136	-0.31	1,555	1.57	844	9,676	1.68
201	266	87	-1.01	26	-1.65	544	159	-1.09
203	77	45	-0.52	2	-1.91	281	11	-1.85
205	4	14	1.10	83	1.82	86	519	1.43
206	2	1	-0.70	4	0.47	7	23	1.08
207	8	9	**0.13**	1	-1.66	54	4	-1.70
208	20	31	0.45	0	-1.91	194	3	-1.94
209	5	42	1.61	0	-1.91	260	1	-1.99
210	9	28	0.99	0	-1.85	172	2	-1.95
212	10	29	0.95	0	-1.99	182	0	-2.00
213	9	34	1.20	0	-1.99	213	0	-2.00

Table 11 – Deposition rates and average concentrations for FPSL test 2-12 (well-mixed)

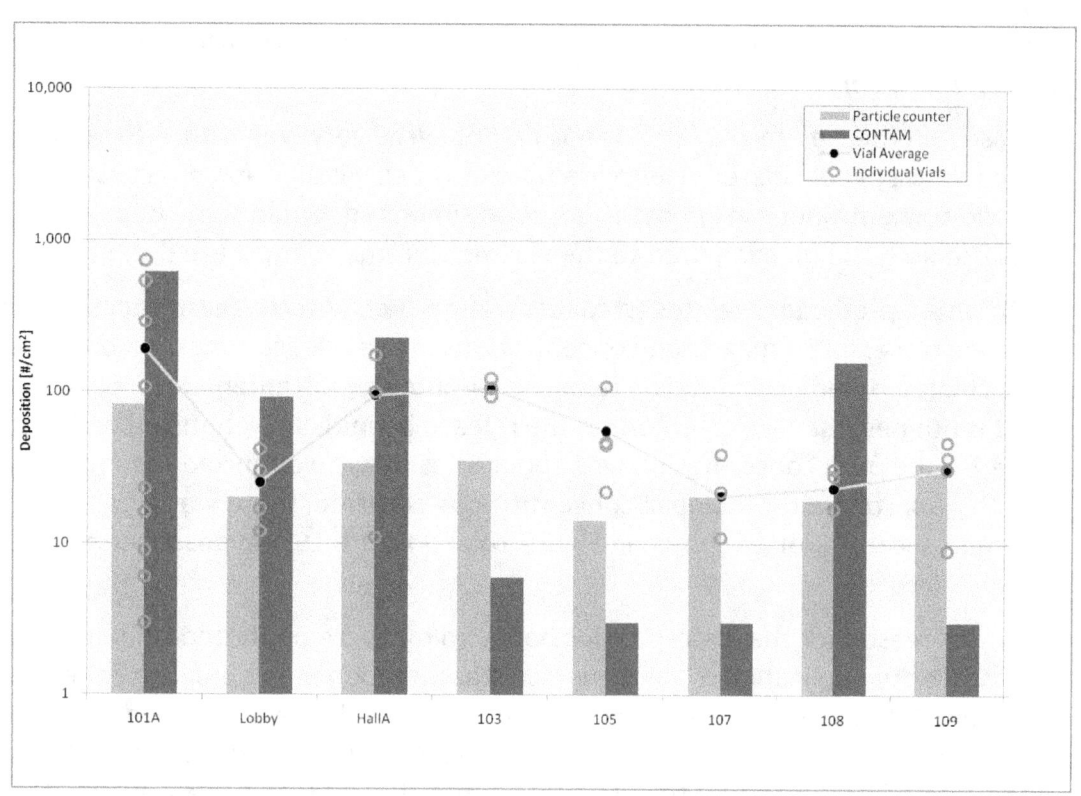

Figure 18 – Deposition for FPSL test 2-6 (well-mixed)

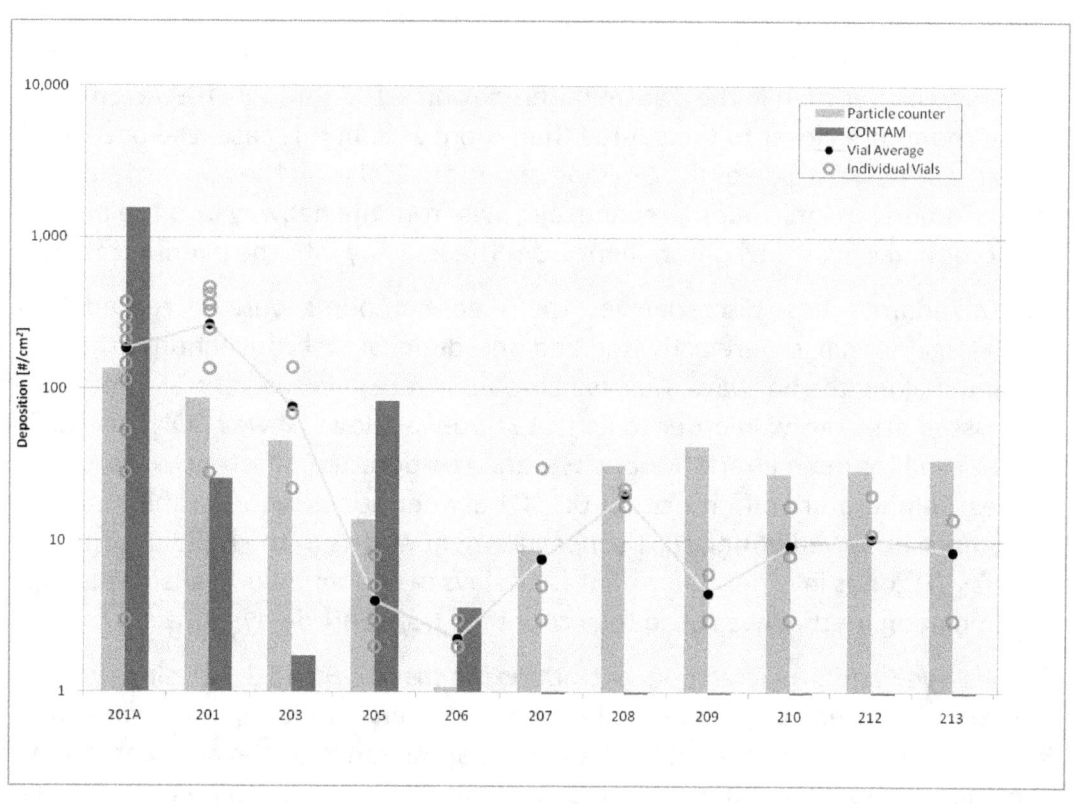

Figure 19 – Deposition for FPSL test 2-12 (well-mixed)

5.2.3 Simulation results for cases 2-6 and 2-12 – 1D convection-diffusion and two-way flow assumptions

The results based on the *well-mixed* simulations do not agree very well with the measured results for the two cases considered. There are transport delay issues and discrepancies in levels of particle concentrations in almost every zone simulated. While some deposition rates are fairly well predicted when compared to the vial measurement, most are not.

One issue with these predictions relates to concentrations within the release zones soon after the releases took place. Initial measured concentrations in the release zones were inconsistent with those calculated based on the stated release amounts and instantaneous mixing within the zone. While it is not necessarily expected that the release be initially well-mixed within the zone, for case 2-6 the peak concentration was about 85 % of the well-mixed assumption and in 2-12 only 50 %. Also, soon after the peak concentrations occurred, there was a significant drop in the measured concentration as shown in Figure 14 and Figure 15 and described in section 5.2.2.

Another potential reason for the discrepancies between measurements and simulations in the well-mixed cases is the application of the *well-mixed* assumption in the building zones. In the two cases discussed, the ventilation systems were not activated. The *well-mixed* assumption is often more reasonable when ventilation systems are operating and mixing is promoted by the system airflows. Another key assumption in the well-mixed cases is the one-way flow through open doorways. In these wind-dominated infiltration cases, one-way flow from the windward to the leeward side of the building can prevent flow reversal through open doorways under prevailing wind conditions. This is evident in Figure 18 where the predicted deposition in room 108 is significantly higher than in the odd-numbered rooms 103 through 109, even though many of these rooms are closer to the source than room 108. In this case, the odd-numbered rooms are on the windward side of the building and room 108 is on the leeward side of the building. The wind-driven infiltration prevents airflows from the hallway into the odd-numbered rooms and drives the contaminant from the hallway into the even-numbered rooms.

In an attempt to address these discrepancies, the release amounts were decreased, two-way flows through large openings were activated and one-dimensional convection-diffusion (1D) was implemented along the hallways. Release amounts were decreased to 25 % of the stated values as discussed previously. In order to activate two-way flows, 2-way CONTAM airflow elements were used for open interior doorways, and temperature differences were established between zones. Temperature differences of 0.1 °C between zones were established as this was a typical difference observed during NIST's measurement efforts in the building. The hallways were defined as 1D zones in order to account for delays in contaminant transport along them, and the 1D simulation method was used to account for transport delays through the ductwork.

Plots of the observed and predicted concentrations for cases 2-6 and 2-12 using these modified assumptions are presented in Figure 20 and Figure 21. When comparing these plots to Figure 14 and Figure 15, it is apparent that the delays in the appearance of elevated contaminant levels are no longer evident and the predicted concentrations appear to be well within an order of magnitude of the measurements in most cases, especially earlier in the test. While these are encouraging observations, the disagreement in peak levels and timing and decay rates shows that inter-zonal airflows are still not being well predicted. One apparent outlier is room 206 in test 2-12 (grey lines at bottom, dashed line appearing between 22:00 and 23:00). This is one of the rooms with a closed door to the hallway. In the simulation, a change in wind direction causes the airflow to reverse direction from outdoors-to-indoors to indoors-to-outdoors, thus drawing air and contaminant into the room from the hallway. While the simulation does not match the measured result very closely, some fluctuations in the concentration are captured. Aside from simulating building pressurization tests, the building model was not adjusted or "tuned" to obtain a more accurate representation of inter-zonal and envelope airflow rates.

Peak magnitude and timing are presented in Figure 22 and Figure 23. Relative to the well-mixed results presented in Figure 16 and Figure 17, the simulated peak values now exhibit a very similar trend to that of the observed values for both cases. The discrepancies in the peak timing also appear to be much improved over the well-mixed case.

Table 12 and Table 13 provide the ASTM guideline parameters. These results are significantly better than those for the well-mixed case provided in Table 8 and Table 9; however, only room 208 in test 2-12 meets all of the recommended levels. The guide indicates that trends in data can be discerned by plotting the residuals versus time. These plots (not presented in this report) revealed similar findings to those revealed by plotting observed and predicted concentrations, e.g., differences in peak levels and timing and decay rates but no systematic trends related to concentration levels.

Deposition rates and average particle concentrations are provided in Table 14 and Table 15 for tests 2-6 and 2-12 respectively. Again, only a few deposition values determined from average particle counter measurements have a *FB* in the ± 0.25 range when compared to the deposition values measured using the vials, but this is not significantly different from the comparison between deposition based on the average particle counter measurements and the vial deposition measurements. However, the measured and predicted average concentrations are in much better agreement as compared to the well-mixed cases (presented in Table 10 and Table 11), with the *FB* of several being within the range of ± 0.25 and most well within an order of magnitude. These observations are revealed by the plots in Figure 24 and Figure 25 (corresponding to Figure 16 and Figure 17 for the well-mixed cases) that show the individual vial measurements, their averages and the average concentrations determined from both the particle counters and CONTAM. In most cases, the deposition based on the average concentrations fell within the range of the individual vial depositions with the variation in deposition amounts between zones being captured fairly well. There does not appear to be a tendency to over- or under-estimate the deposition vial measurements for either the measured or predicted deposition values. This together with the values being within the range of vial deposition measurements tends to indicate that the settling velocity used as the deposition rate for the 1 μm particles is reasonable.

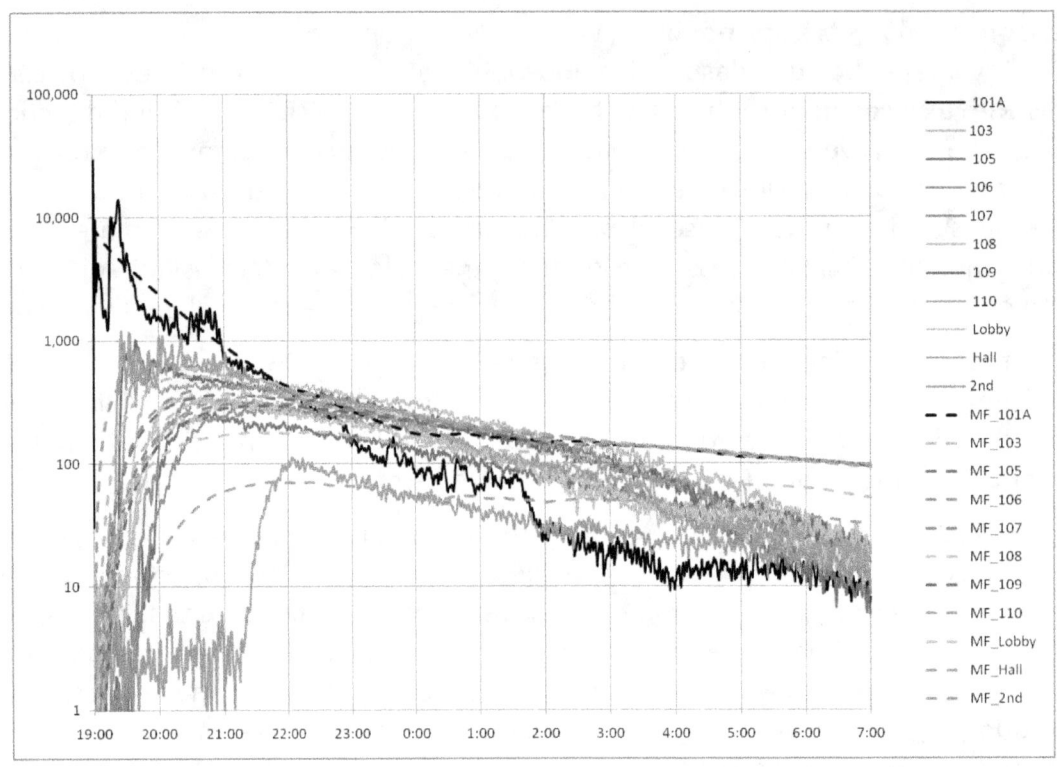

Figure 20 – Predicted (2-way, 1D, 25 % Release) vs. measured concentrations for FPSL test 2-6

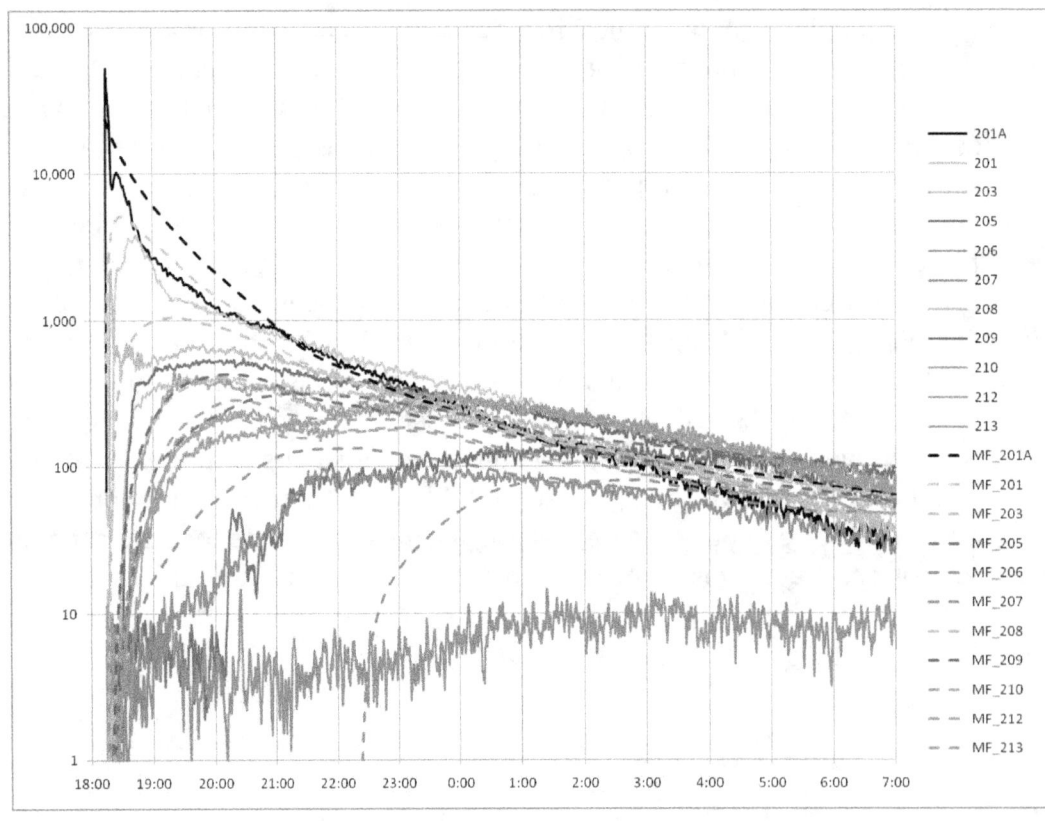

Figure 21 – Predicted (2-way, 1D, 25 % Release) vs. measured concentrations for FPSL test 2-12

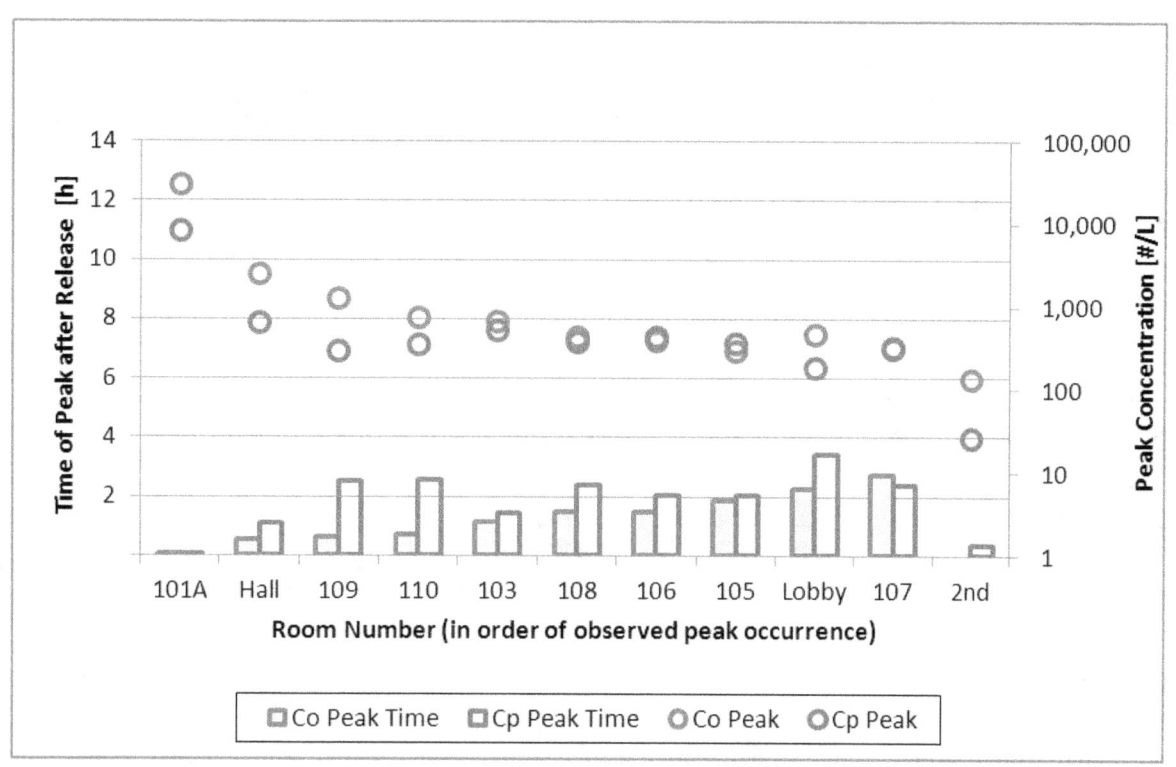

Figure 22 – Peak timing for FPSL test 2-6 (2-way, 1D, 25 % Release case)

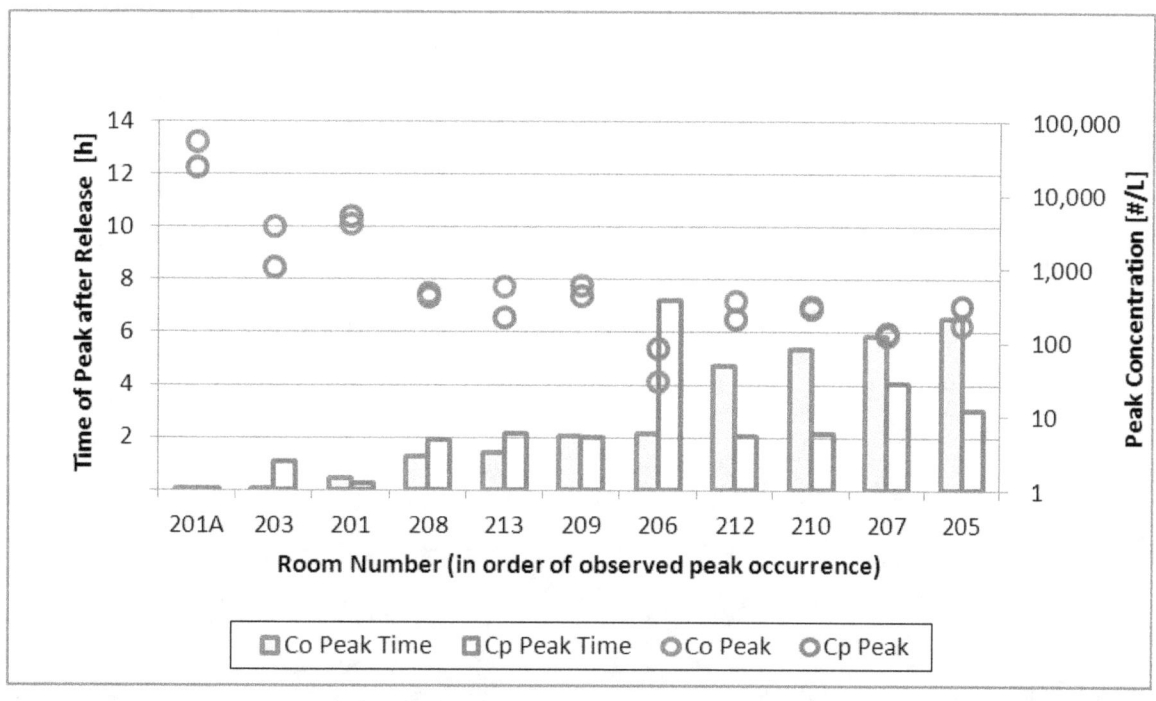

Figure 23 – Peak timing for FPSL test 2-12 (2-way, 1D, 25 % Release case)

	Criteria	101A	Lobby	Hall	103	105	106	107	108	109	110	2nd
N		4321	4321	4321	4321	4321	4321	4321	4321	4321	4321	4321
r	≥ 0.9	0.74	**0.91**	0.88	**0.96**	**0.96**	**0.97**	**0.93**	**0.92**	0.50	0.74	0.69
m	0.75 - 1.25	0.61	0.38	0.60	0.71	**1.20**	**1.03**	**0.79**	**0.85**	0.19	0.44	0.44
$b/\overline{Co}$	≤ 0.25	0.63	0.49	0.59	0.30	0.90	0.68	0.56	0.72	0.58	0.50	1.02
$NMSE$	≤ 0.25	3.17	0.35	0.32	**0.08**	0.63	0.32	**0.14**	0.27	0.82	0.31	0.39
$|FB|$	≤ 0.25	**0.21**	-0.15	0.17	**0.00**	0.71	0.52	0.30	0.44	-0.26	**-0.06**	0.37
$|FS|$	≤ 0.50	**-0.39**	-1.41	-0.74	-0.60	**0.45**	**0.12**	**-0.33**	**-0.16**	-1.48	-0.97	-0.85

Table 12 – ASTM guide statistics for FPSL test 2-6 (2-way, 1D, 25 % Release)

	Criteria	201A	201	203	205	206	207	208	209	210	212	213
N		4591	4591	4591	4591	4591	4591	4591	4591	4591	4591	4591
R	≥ 0.9	0.85	**0.92**	0.78	-0.02	0.54	0.79	**0.94**	**0.96**	0.59	0.83	0.88
M	0.75 - 1.25	**1.02**	1.50	**1.00**	-0.04	4.59	**0.92**	**0.83**	0.71	0.61	0.55	0.50
$b/\overline{Co}$	≤ 0.25	0.49	-0.28	**0.15**	2.14	1.20	0.53	**0.09**	**-0.01**	**0.22**	**0.09**	**0.05**
$NMSE$	≤ 0.25	2.61	0.87	0.41	1.14	7.35	**0.24**	**0.06**	**0.19**	**0.21**	0.31	0.49
$|FB|$	≤ 0.25	0.41	**0.20**	**0.14**	0.71	1.41	0.37	**-0.08**	-0.34	**-0.18**	-0.43	-0.58
$|FS|$	≤ 0.50	**0.36**	0.91	**0.49**	1.12	1.95	**0.32**	**-0.26**	-0.57	**0.07**	-0.78	-1.01

Table 13– ASTM guide statistics for FPSL test 2-12 (2-way, 1D, 25 % Release)

Room	Deposition [#/cm²]					Average Concentration [#/L]		
	Vial Average	Particle Counter	Fractional Bias	CONTAM	Fractional Bias	Particle Counter	CONTAM	Fractional Bias
101A	192	82	-0.80	85	-0.77	543	671	**0.21**
Lobby	25	20	**-0.22**	15	-0.53	134	115	**-0.15**
Hall	95	34	-0.96	30	-1.03	222	264	**0.17**
103	104	35	-1.00	29	-1.12	231	232	**0.00**
105	56	14	-1.18	25	-0.76	95	198	0.71
106	n/a	19	n/a	28	n/a	127	218	0.52
107	21	20	**-0.02**	23	**0.10**	134	181	0.30
108	23	19	**-0.18**	25	**0.08**	128	200	0.44
109	31	34	**0.08**	22	-0.35	223	172	-0.26
110	n/a	30	n/a	24	n/a	201	189	**-0.06**
2nd	n/a	5	n/a	1	n/a	32	10	-1.07

Table 14 – Deposition rates and average concentrations for FPSL test 2-6 (2-way, 1D, 25 % Release)

Room	Deposition [#/cm²]					Average Concentration [#/L]		
	Vial Average	Particle Counter	Fractional Bias	CONTAM	Fractional Bias	Particle Counter	CONTAM	Fractional Bias
201A	185	136	-0.31	171	**-0.08**	844	1,277	0.41
201	266	87	-1.01	89	-0.99	544	667	**0.20**
203	77	45	-0.52	43	-0.56	281	322	**0.14**
205	4	14	1.10	24	1.43	86	180	0.71
206	2	1	-0.70	5	0.80	7	39	1.41
207	8	9	**0.13**	11	0.32	54	79	0.37
208	20	31	0.45	24	**0.20**	194	178	**-0.08**
209	5	42	1.61	25	1.38	260	184	-0.34
210	9	28	0.99	19	0.69	172	143	**-0.18**
212	10	29	0.95	16	0.41	182	117	-0.43
213	9	34	1.20	16	0.60	213	118	-0.58

Table 15 – Deposition rates and average concentrations for FPSL test 2-12 (2-way, 1D, 25 % Release)

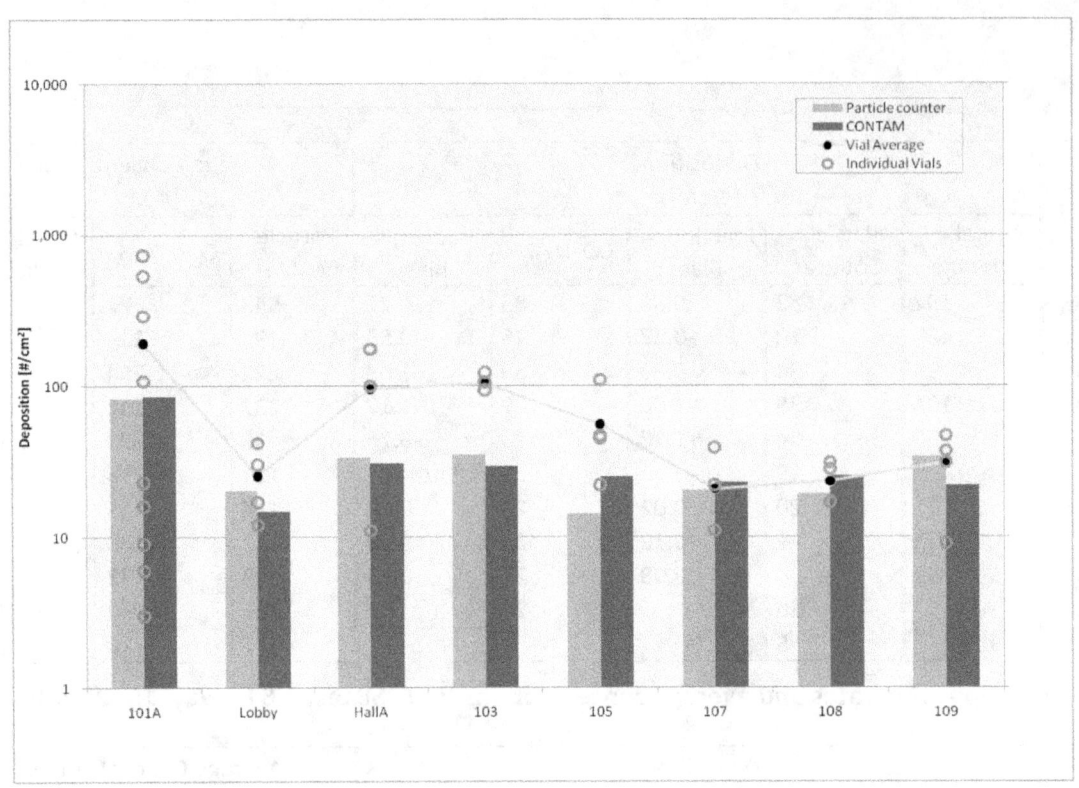

Figure 24 – Deposition for FPSL test 2-6 (2-way, 1D, 25 % release)

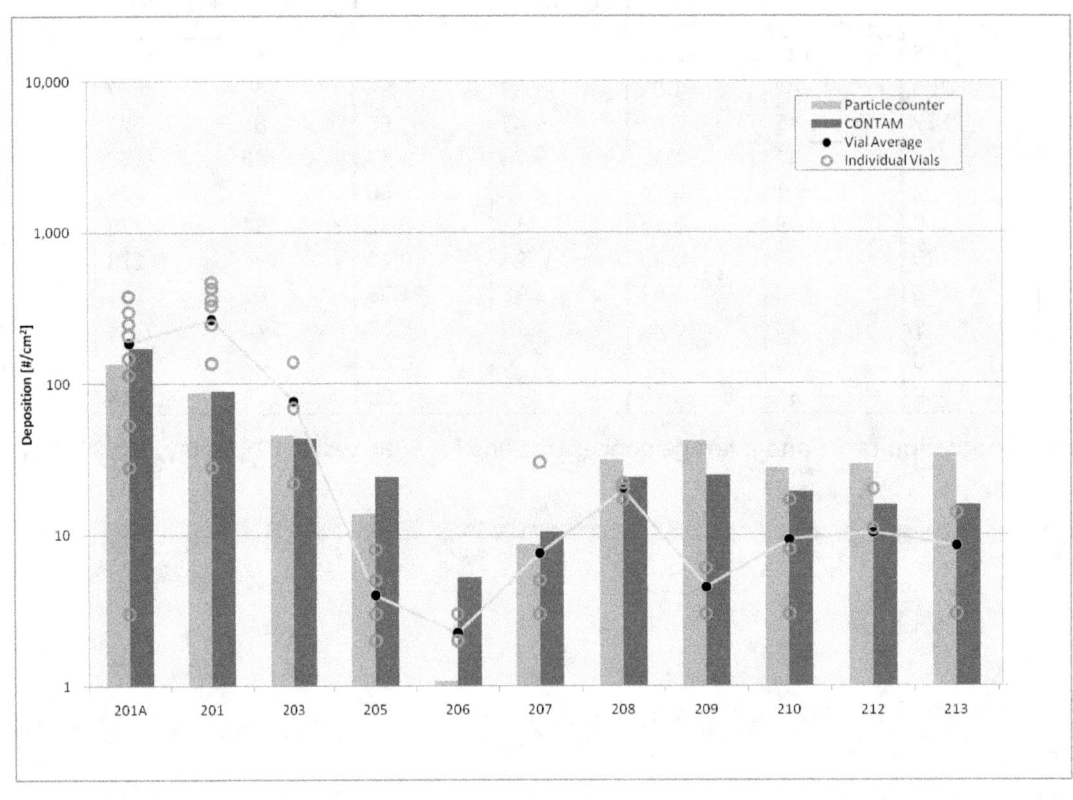

Figure 25 – Deposition for FPSL test 2-12 (2-way, 1D, 25 % release)

5.2.4 Comparison of average concentrations (for remaining test cases)

The characterization cases that were accompanied by vial deposition measurements (2-6 and 2-12) were used to establish the simulation method for the remainder of the cases for which vial deposition was not measured directly. For these cases, only the ASTM guide values and average concentrations between measured and simulation are compared. Averaging periods began upon release and continued until the so-called purge cycle between events, during which windows were opened and the air handlers were activated. For all but two of the remaining cases (1-4 and 1-10), the air handler on the release floor was turned on for approximately thirty seconds to promote mixing of the release agent. Summaries of the characterization test configurations evaluated for this study are presented in Table 16 and Table 17 including release locations and stated release amounts. Detailed results are presented in the Appendices. Simulated release amounts are provided in the titles of the charts presented in the appendices. The simulated FPSL releases were all set to 25 % of the stated release amounts with the exception of test 2-8 which was set to 2 % to better match the source location measurements after a very brief initial spike in concentration. However, in the visolite release cases, the release efficiency was reduced significantly to about 1 % of the stated release amounts. In two of the cases, 1-5 and 1-6, the release was simulated as a timed release (as opposed to the burst source used for all other simulations) in order to better match the gradual buildup in concentration displayed in the release zone for these two tests.

Appendix A contains log-scale plots of observed versus predicted concentration for both sets of characterization tests, INL-1 and INL-2. Appendix B contains bar charts of observed versus predicted average concentrations followed by tabulated values. The bar charts are provided in log-scale in order to provide relative order-of-magnitude comparisons as agreement in the trends are more reasonable to obtain than absolute agreement. Appendix C provides the ASTM guide results and Appendix D provides plots of the weather data during both sets of characterization tests.

Comparisons between observed and predicted values based on the ASTM guide reveal very few instances in which acceptable agreement is obtained. However, it is encouraging to note that there are cases in which values are in the acceptable range, as depicted in the charts in Appendix B and tables in Appendix C, many of the predicted average values are within an order of magnitude of the measured average values. Further examination of the results does not reveal any distinct trends between predictability and zone properties, e.g., proximity to the source, door open status, etc. However, the building model is also able to capture observed variations in contaminant levels that were likely due to variations in wind speed and direction that occurred in many of the tests including the fairly obvious opening of exterior windows during the purge cycles. Note that these periods were not used in the comparison of average concentrations due to the uncertainty in airflows through the open windows. Although a rigorous sensitivity analysis was not carried out, it was noted that decay rates are very sensitive to wind pressures as revealed in the concentration plots for tests 1-4 and 2-6 in Appendix A and the ASTM guide results provided in Table 36 and Table 37 in Appendix C. As previously discussed, it was assumed that surface averaged wind pressures were based on a *typical* rectangular building, but the tent around the building significantly challenges this assumption.

35

ID	Date	Time	Material	Amount	Location	HVAC	Windows	Doors Closed
1	21-Aug-07	11:35	FPSL	125 mg	Lobby	On at 12:17	14:00 open	none
2	21-Aug-07	15:10	"	125 mg	"	On for 30 s, filters ON Intakes	15:53 open	"
3	21-Aug-07	16:25	"	125 mg	"	On for 30 s, filters OFF Intakes	16:57 open	"
4	21-Aug-07	17:45	"	125 mg	"	OFF, lobby fan On 45 s after release	Closed	"
5	22-Aug-07	8:35	"	125 mg	"	On for 30 s	9:16 open	"
6	22-Aug-07	10:05	"	125 mg	"	On for 30 s	10:50 open	109, 110
7	22-Aug-07	11:50	Visolite	1000 mg	"	On for 30 s	12:59 open	"
8	22-Aug-07	13:59	"	500 mg	"	On for 30 s	14:36 open	"
9	22-Aug-07	16:42	FPSL	125 mg	201	On for 30 s	17:18 open	212, 213
10	22-Aug-07	17:57	Visolite	1000 mg	"	On for 30 s	Closed	none
11	23-Aug-07	8:47	"	500 mg	"	On for 30 s	9:21 open	205, 206, 207
12	23-Aug-07	10:02	"	1000 mg	"	On for 30 s	10:40 open	"
13	23-Aug-07	11:30	"	1000 mg	"	OFF	12:06 open	"

Table 16 – Test conditions for INL-1 characterization

ID	Date	Time	Material	Amount	Location	HVAC	Windows	Doors Closed
1	18-Aug-08	9:00	FPSL	10 mg	101	On at 9:40	9:50 open	Lob-101, 101-103
2	18-Aug-08	11:55	"	1 mg	101	On at 12:35	12:35 open	"
3	18-Aug-08	13:40	"	1 mg	110	On at 14:10	14:10 open	"
4	18-Aug-08	15:25	"	1 mg	101A	On at 16:05	16:05 open	"
5	18-Aug-08	17:00	"	1 mg	101A	On at 17:40	17:40 open	"
6	18-Aug-08	19:00	"	1 mg	101A	OFF	Closed	"
7	19-Aug-08	8:20	"	1 mg	101A	On at 9:00	9:00 open	"
8	19-Aug-08	10:05	"	1 mg	201A	On at 10:45	10:45 open	205, 206, 207
9	19-Aug-08	12:00	"	2 mg	201A	On at 13:00	13:00 open	"
10	19-Aug-08	14:15	"	1 mg	201A	On at 15:05	15:05 open	"
11	19-Aug-08	16:10	"	1 mg	201A	On at 16:50	16:50 open	"
12	19-Aug-08	18:15	"	2 mg	201A	OFF	Closed	"

Table 17 – Test conditions for INL-2 characterization

6 Summary and Conclusions

A multizone building model validation study was performed in a small two-story office building in which a number of particle release experiments were conducted. This study was undertaken to determine the usefulness of utilizing multizone modeling as a means to simulate internal release scenarios to support and enhance sample planning methods. The release experiments that were carried out in the building were part of sample planning exercises that are quite resource intensive. Building simulations have the potential to allow virtual release experiments to be performed at a much reduced level-of-effort than that required by actual experiments.

Inputs to the building model were based on design documents and as-built measurements. Building envelope leakage characteristics were obtained by building pressurization tests, and ventilation system airflows were measured directly. Particle deposition rates were assumed to be constant and uniform, based on the particle settling velocity of a 1 μm particle. Most tests were performed with the ventilation system off, so wind was the dominant driving force for airflow. However, due to a very large decontamination tent that surrounded the building, the wind effects were one of the most uncertain inputs to the building model.

Two rounds of releases were performed during the summers of 2007 and 2008. This study focused on the so-called *building characterization* releases that were performed prior to the actual field evaluation studies. Field evaluation studies refer to the release events wherein a biological simulant was released and sampling strategies and sampling methods were evaluated with decontamination performed between events. The *characterization* tests were performed with non-biological aerosol releases in order to establish release scenarios that would lead to gradients in deposition amounts across individual floors of the building. During these tests, real-time particle aerosol monitors were used to measure particle concentrations in about ten locations on the floor being tested.

Comparisons between observed and simulated results were based upon the ASTM D5157 *Standard Guide for the Evaluation of Indoor Air Quality Models*. This standard includes a set of statistical indicators and recommended ranges of values into which the indicators should fall, as well as other qualitative observations of side-by-side comparisons. Comparisons between average observed and predicted concentrations were emphasized as indicators of agreement between deposition amounts wherein the Fractional Bias (FB) indicates the level of agreement, i.e., the absolute value of the FB being less than 1.64 indicates that the values agree within an order-of-magnitude.

An initial set of simulations was performed on two release cases during which deposition was measured directly using settling vials. Initially, the building was simulated using the well-mixed assumption in all zones, one-way flows between zones and one hundred percent release efficiency of the release agent. Based on the results of these initial simulations, several modeling assumptions were modified and then employed in the remaining simulations including: reducing the release efficiency to 25 % to as low as 1 % depending on the release agent and mechanism, implementing one-dimensional convection-diffusion transport in the hallways and ducts (as opposed to the well-mixed assumption), and implementing two-way flow through open doorways. These modifications in the simulation method provided much better agreement than did the well-mixed method.

Comparisons based on the ASTM guide method revealed very few cases in which recommended level of agreement were obtained. However, the results were encouraging when the data were evaluated for trends between measured and predicted values. Qualitative evaluation of contaminant time histories revealed relative magnitudes in peak levels and timing were fairly well captured in many cases. Trends in average concentrations across the floor under test were fairly well captured as seen in bar charts provided in Appendix B, and fractional bias in these values indicate they were almost always within an order-of-magnitude.

The experiments were not performed for the purposes of a model validation study, which limits the extensiveness of the conclusions. Nevertheless, under the circumstances, the modeling performed fairly well from a qualitative standpoint in that gradients across a given floor of test were fairly well captured. Therefore, multizone modeling can be expected to prove useful in providing insight for developing sampling scenarios and experiment designs and for increasing the general understanding of building behavior under various release scenarios.

This study reflects the importance of obtaining as-built building characteristics when trying to develop building airflow and contaminant transport simulation models. These models can be sensitive to inputs, for example, as revealed by the nature of wind pressure effects on the building envelope and thus the internal flows between building zones and ultimately contaminant transport characteristics. Future work should focus on establishing better validation test cases. Such cases should involve more realistic building configurations, better characterized set of interzone airflow properties and pressure differences, and improved understanding of release mechanisms and particle behavior to ensure particle-particle and particle-surface effects are being accounted for. Modeling approaches could also be modified to include sensitivity analysis on model inputs and utilizing CFD capabilities of combined multi-zone/CFD modeling within a select set of interior zones and or evaluating building pressures via external CFD modeling.

Acknowledgements

This report was supported by the U.S. Department of Homeland Security, Science and Technology Directorate, under IAA # HSHQDC-08-X-00410, with the support of Dr. Bert Coursey, Director of the Office of Standards, Department of Homeland Security and of the NIST Office of Law Enforcement Standards. The authors would like to express their appreciation to the members of the Validated Sampling Plan Working Group led by Dr. Randy Long of the Department of Homeland Security. We would also like to express our thanks to Dr. David Silcott and Tom Pilholski of ICx Technologies, Inc. for providing and helping us to gain an understanding of the *building characterization* data.

References

1.	Sanderson, W.T., et al., *Surface Sampling Methods for Bacillus anthracis Spore Contamination.* Emerging Infectious Diseases, 2002. **8**(10): p. 1145-1151.

2.	Rose, L., et al., *Swab Materials and Bacillus anthracis Spore Recovery from Nonporous Surfaces.* Emerging Infectious Diseases, 2004. **10**(6): p. 1023-1029.

3.	*Evaluation Report September 2007: Indoor Field Evaluation of Sample Collection Methods and Strategies at Idaho National Laboratory.* 2008, Department of Homeland Security and Joint Program Executive Office for Chemical and Biological Defense: Washington, D.C.

4.	*Evaluation Report September 2008: Indoor Field Evaluation of Sample Collection Methods and Strategies at Idaho National Laboratory II.* 2009, JPEO-CBD.

5.	Walton, G.N. and W.S. Dols, *CONTAM 2.4 User Guide and Program Documentation.* 2005, National Institute of Standards and Technology: Gaithersburg.

6.	Matzke, B.D., et al., *Visual Sample Plan Version 5.0 User's Guide.* 2007, Pacific Northwest National Laboratory: Richland, WA.

7.	Dols, W.S., et al., *Development and Demonstration of a Method to Evaluate Bio-Sampling Strategies using Building Simulation and Sample Planning Software.* 2009, National Institute of Standards and Technology: Gaithersburg.

8.	ASTM, *Standard Test Method for Determining Air Leakage Rate by Fan Pressurization.* 2003, American Society for Testing and Materials: USA.

9.	Emmerich, S.J. and A.K. Persily, *Airtightness of Commercial Buildings in the U.S,* in *26th Air Infiltration and Ventilation Centre Conference.* 2005, AIVC: Brussels.

10.	Brennan, T., et al., *Measuring Airtightness at ASHRAE Headquarters.* ASHRAE Journal, 2007. **49**(September).

11.	ATTMA, *Measuring Air Permeability of Building Envelopes.* 2007, The Air Tightness Testing and Measurement Association.

12.	Emmerich, S.J., et al., *Air and Pollutant Transport from Attached Garages to Residential Living Spaces.* 2003, National Institute of Standards and Technology: Gaithersburg, MD.

13.	Persily, A.K. and E.M. Ivy, *Input Data for Multizone Airflow and IAQ Analysis.* 2001, National Institute of Standards and Technology: Gaithersburg.

14.	*IBAC Real Time Biological Aerosol Monitor.* March 2010]; Available from: http://www.icxt.com/products/icx-detection/biological/ibac/.

15.	ASHRAE, *Fundamentals Handbook.* 2009: Atlanta.

16.	Zhang, Y., *Indoor Air Quality Engineering.* 2005, Boca Raton: CRC Press.

17.	Lai, A.C.K. and W.W. Nazaroff, *MODELING INDOOR PARTICLE DEPOSITION FROM TURBULENT FLOW ONTO SMOOTH SURFACES.* Journal of Aerosol Science, 1999. **31**(4): p. 463-476.

18.	ASTM, *Standard Guide for the Evaluation of Indoor Air Quality Models.* 2003, American Society for Testing and Materials.

Appendix A – Measured vs. Predicted Concentration

This appendix provides plots of measured versus predicted concentrations for all building characterization test cases for both the INL-1 and INL-2 sets of experiments. All values are in units of #/L (number of particles per liter), and the measured values are plotted as 100-minute running averages in order to smooth the data. Simulations were performed employing the 1D convection-diffusion, two-way openings and reduced release amounts as provided in the figure titles. In these plots the measured values are represented by solid lines, the predicted values by the dashed lines and the room wherein the source was located by the black lines.

INL-1 Characterization Tests

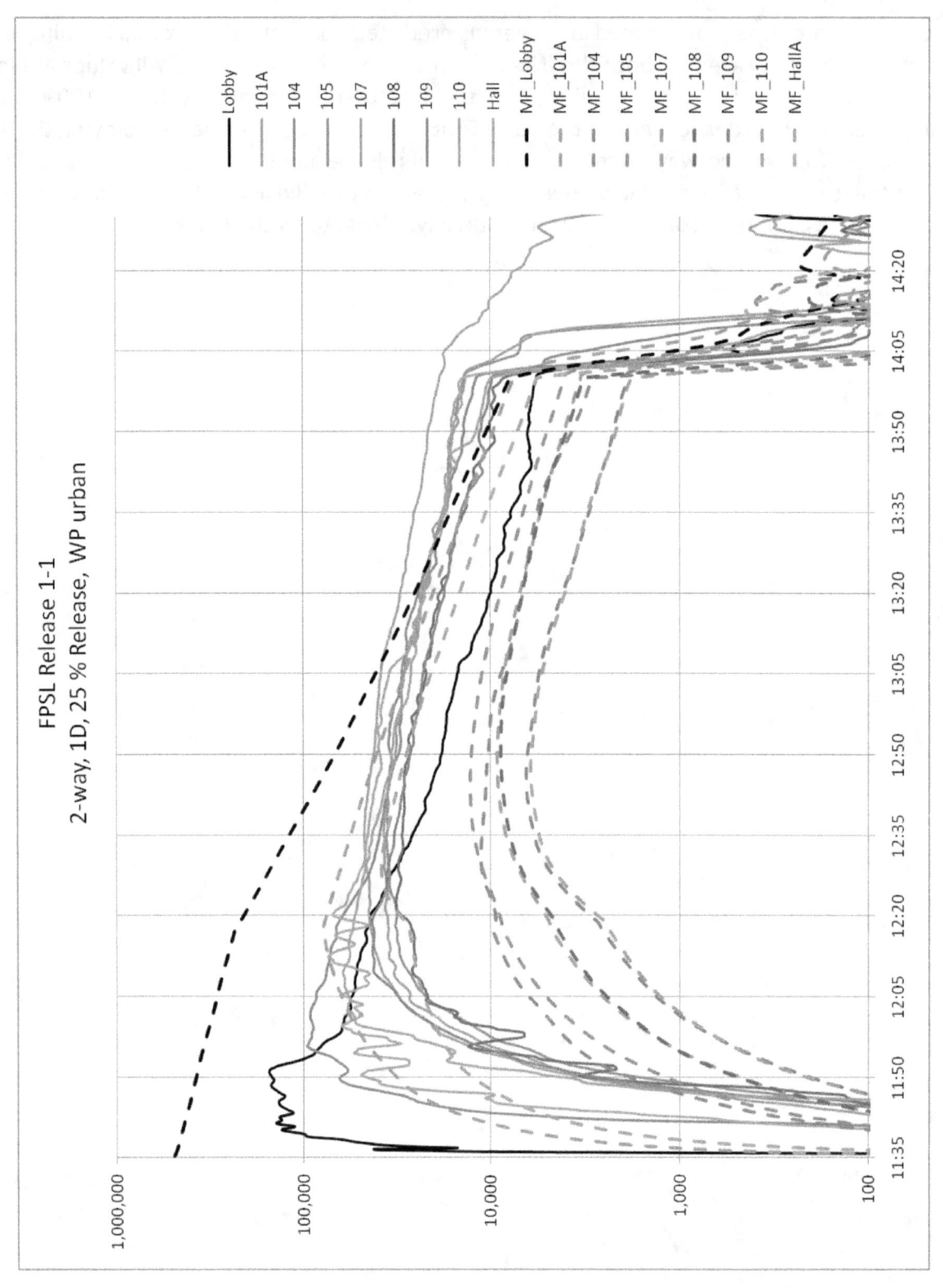

FPSL Release 1-1
2-way, 1D, 25 % Release, WP urban

Figure 26 – Test 1-1

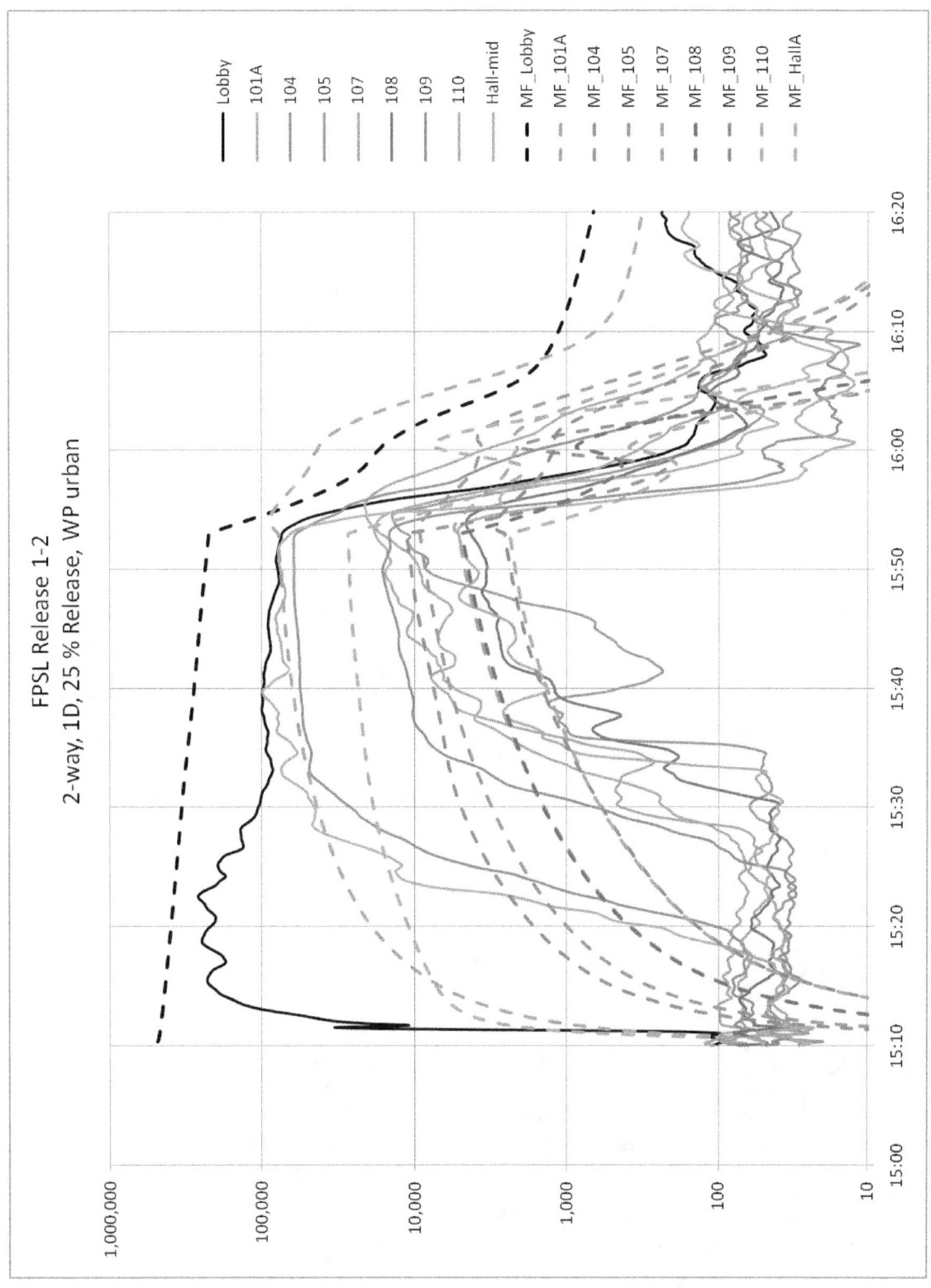

Figure 27 – Test 1-2

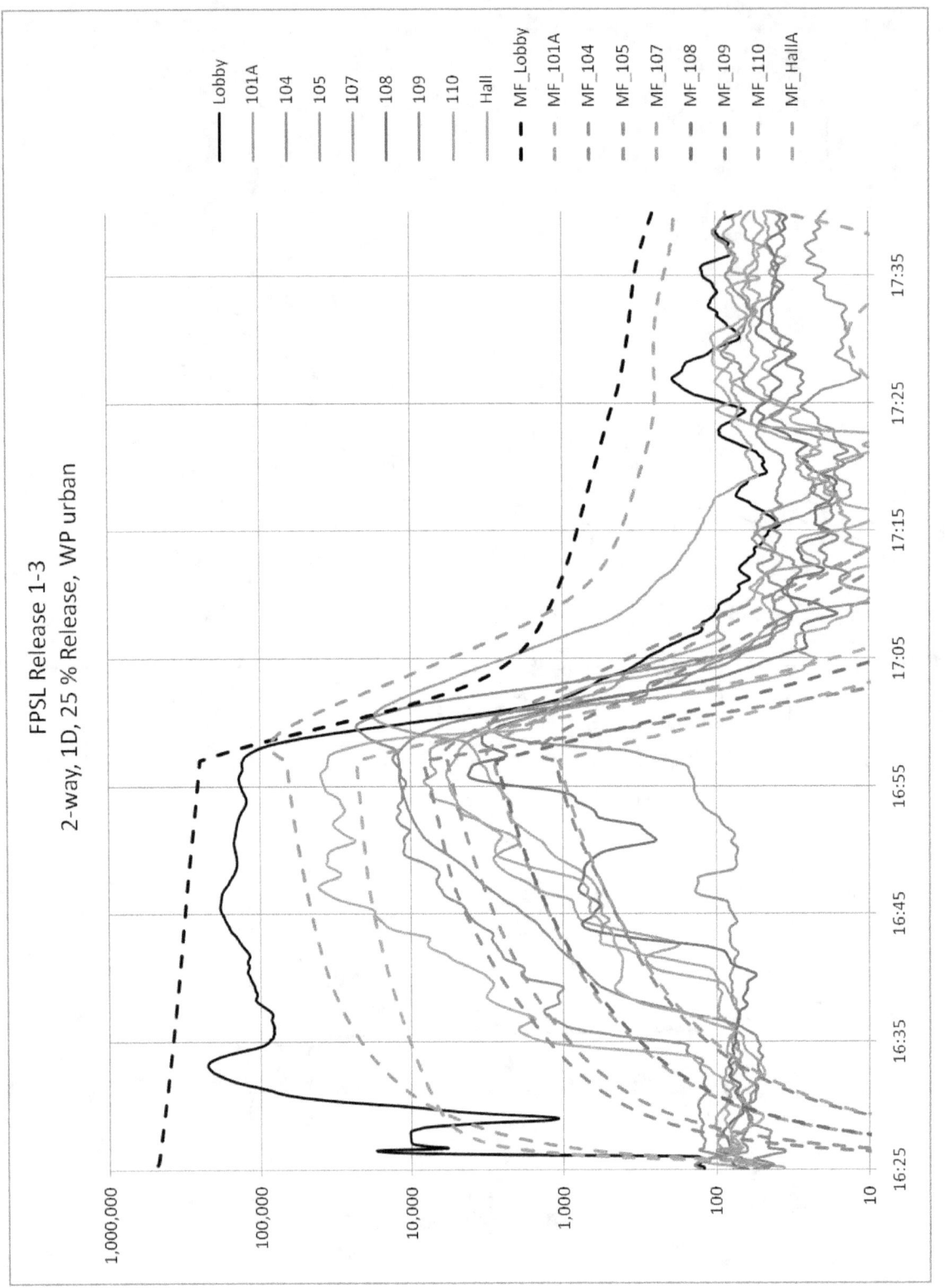

Figure 28 – Test 1-3

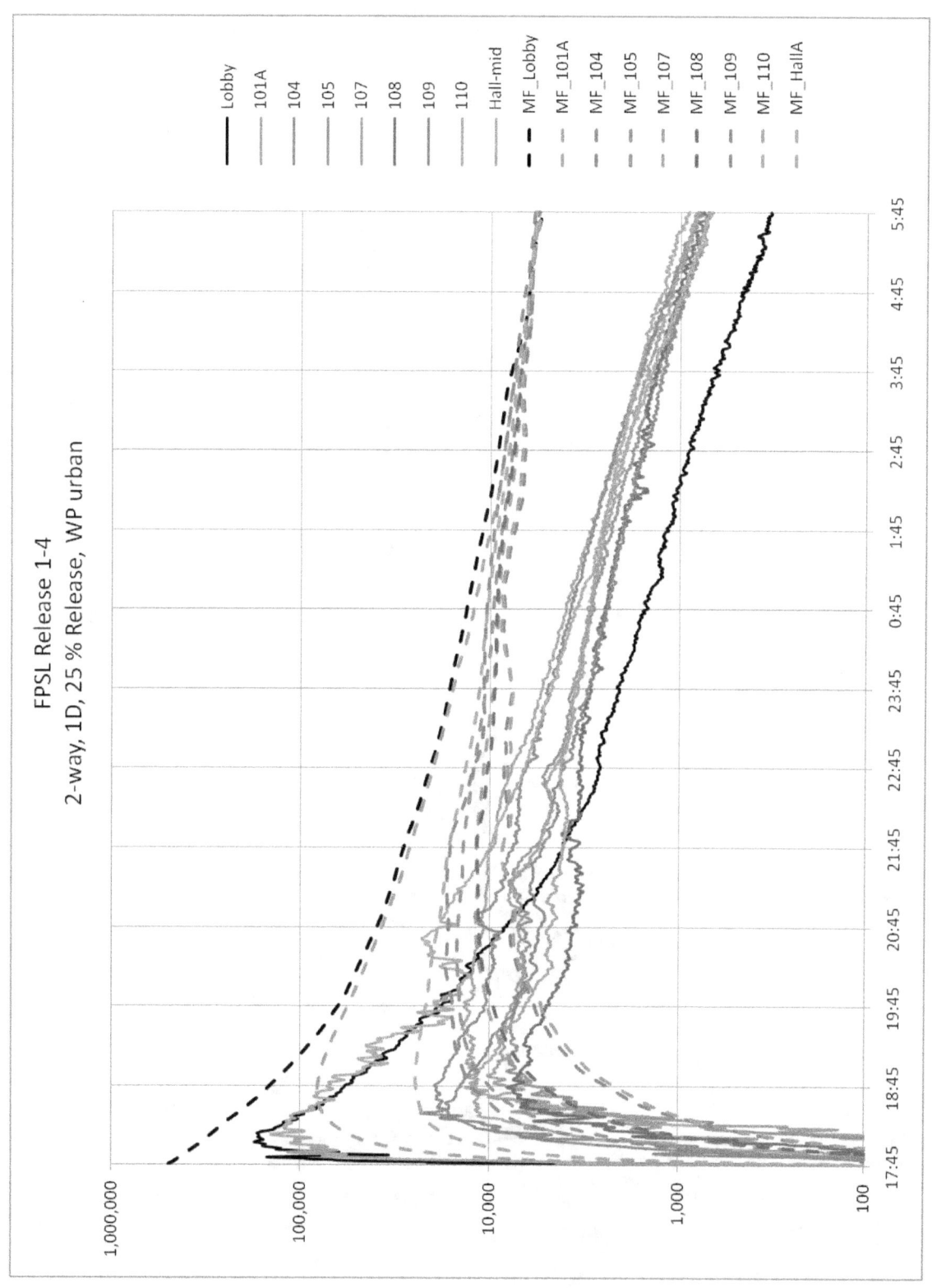

Figure 29 – Test 1-4 (urban wind pressure terrain effect)

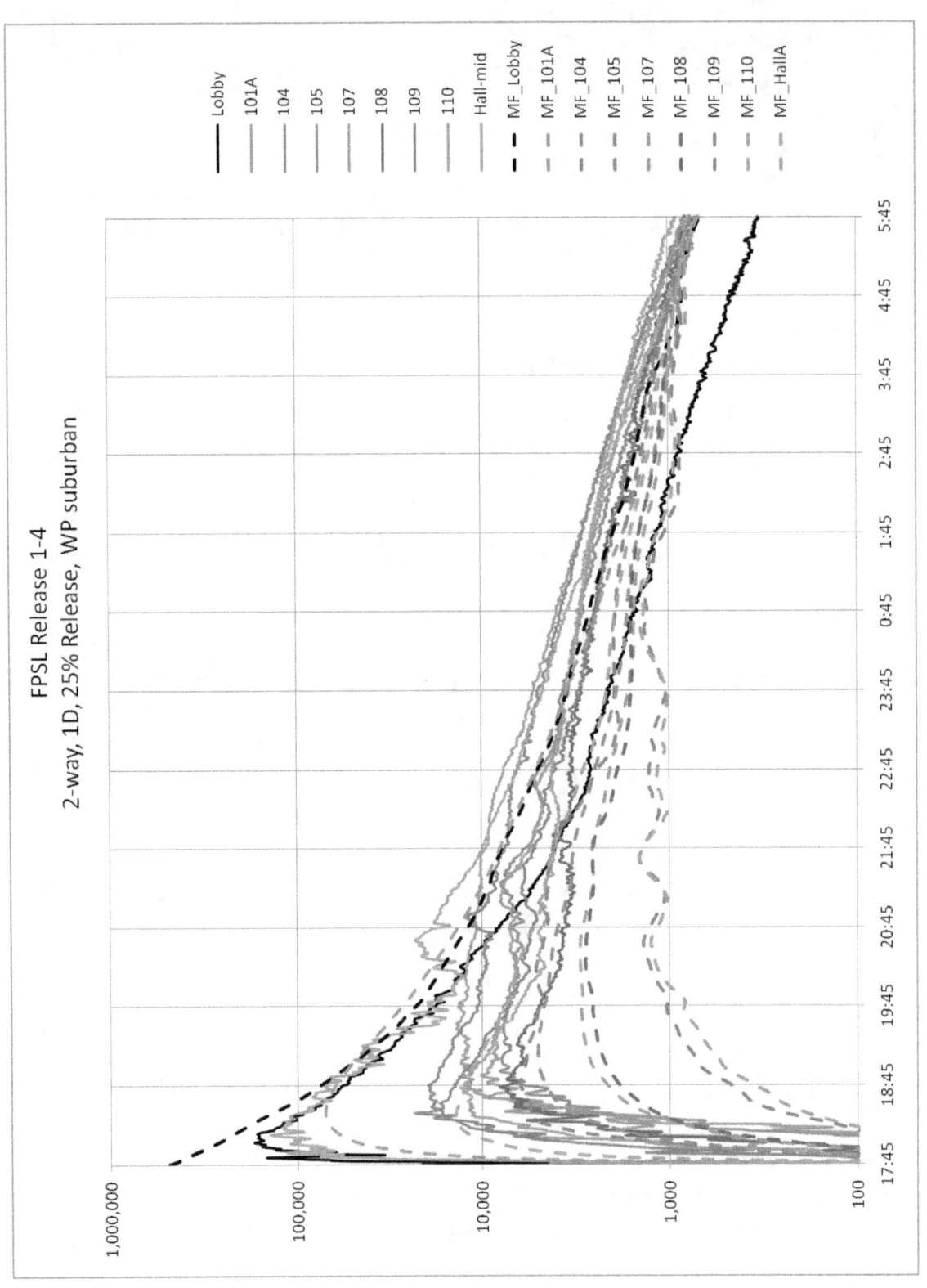

Figure 30 – Test 1-4 (suburban wind pressure terrain effect)

46

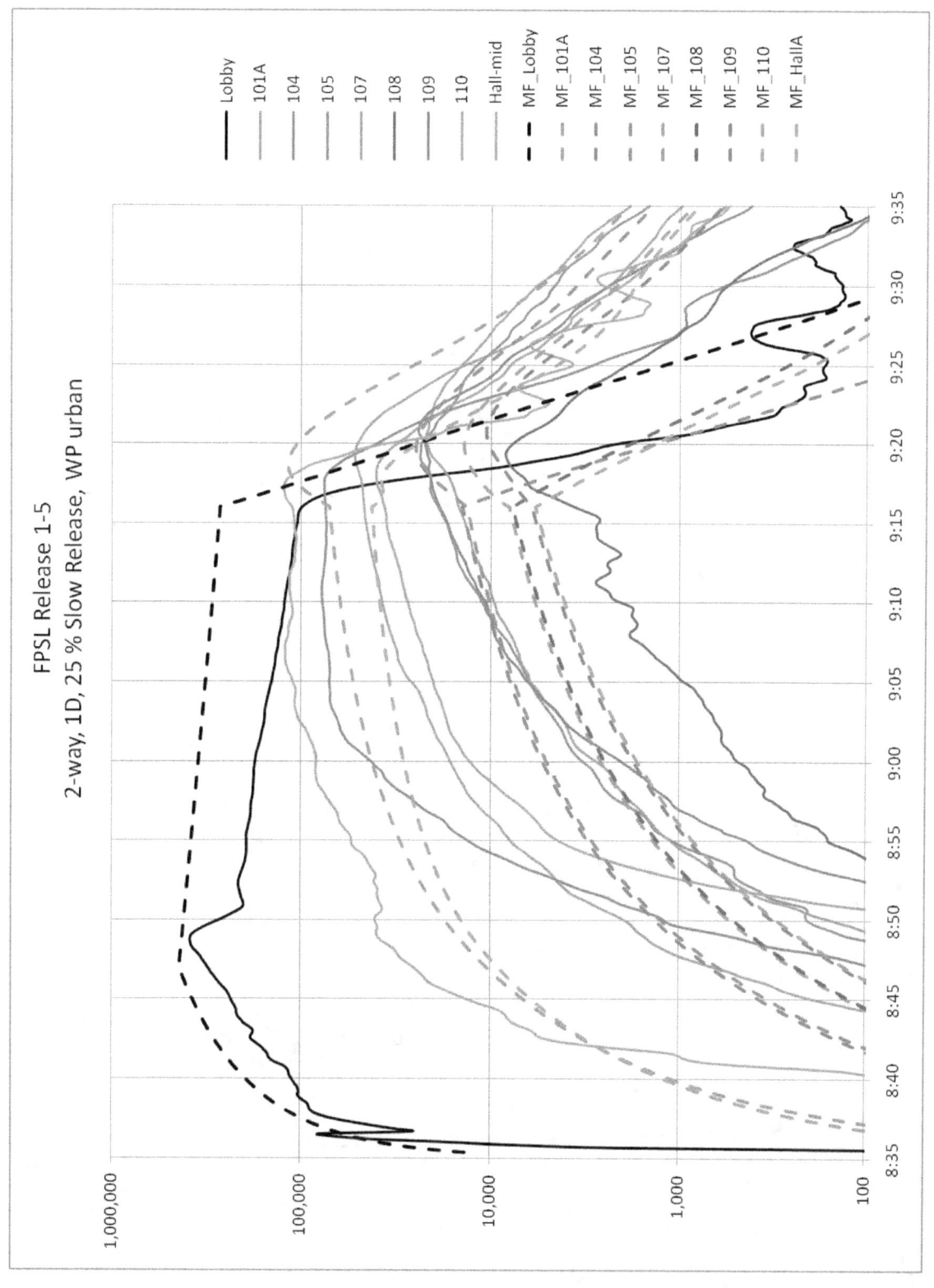

Figure 31 – Test 1-5

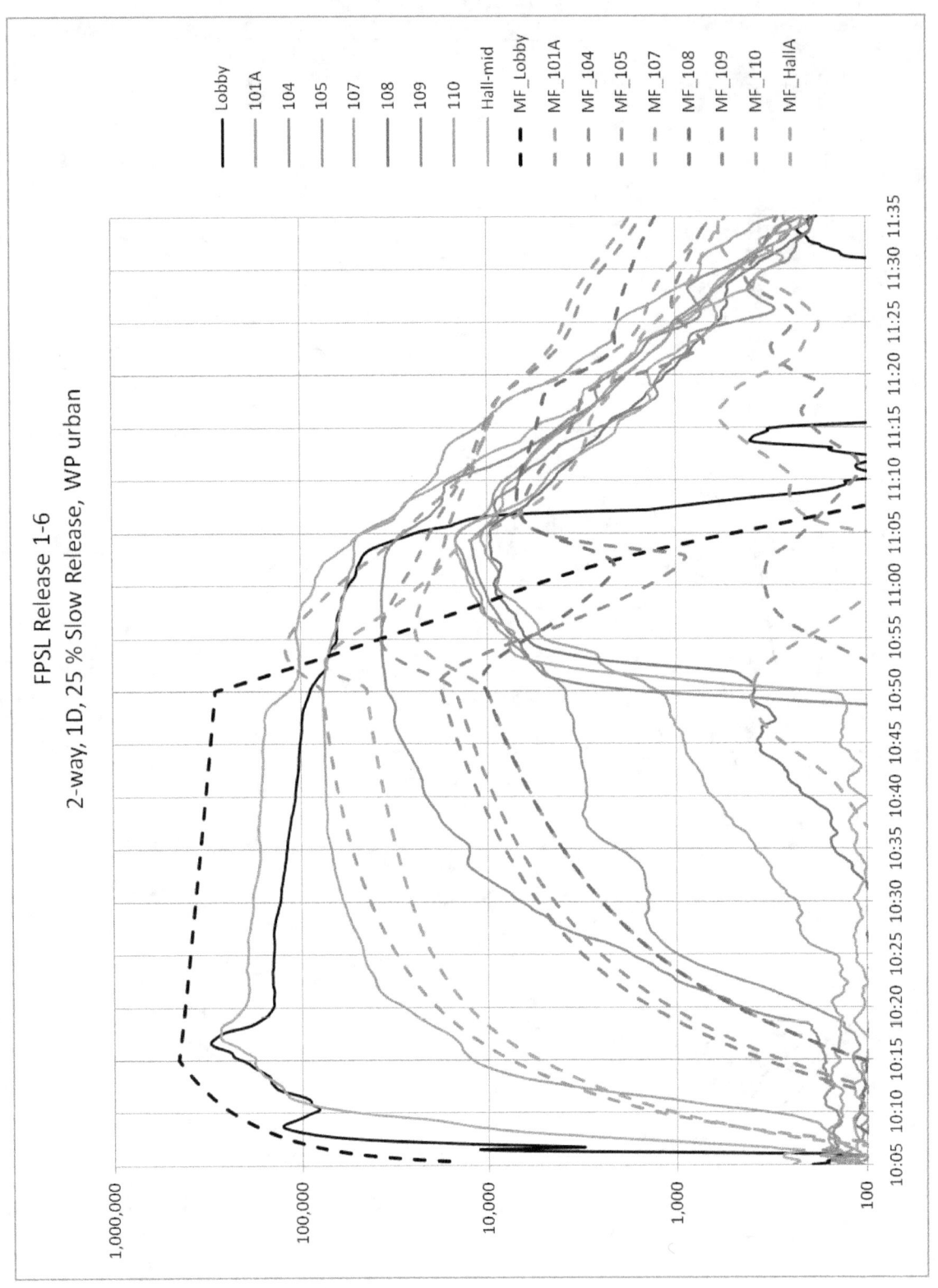

Figure 32 – Test 1-6

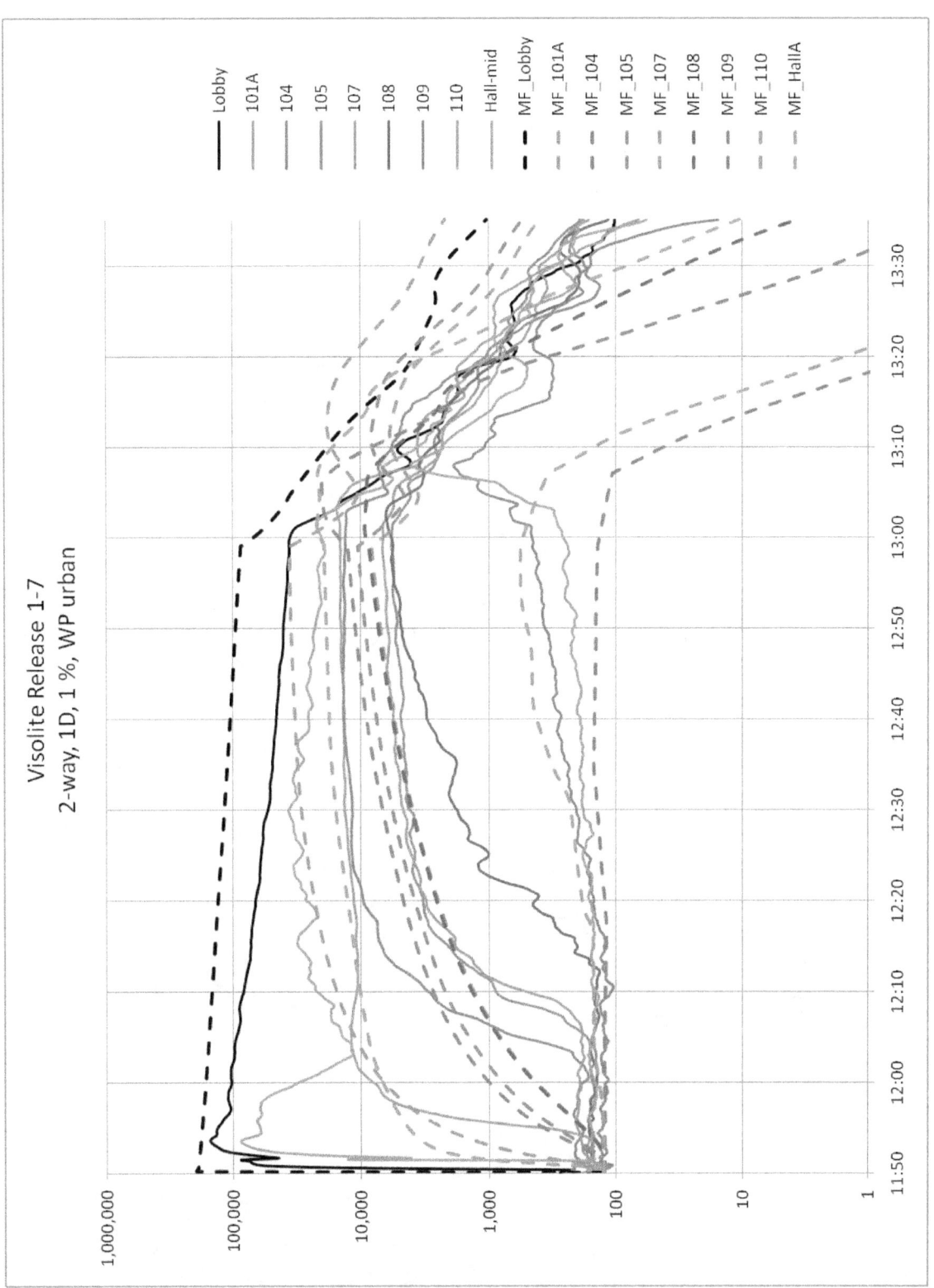

Figure 33 – Test 1-7

49

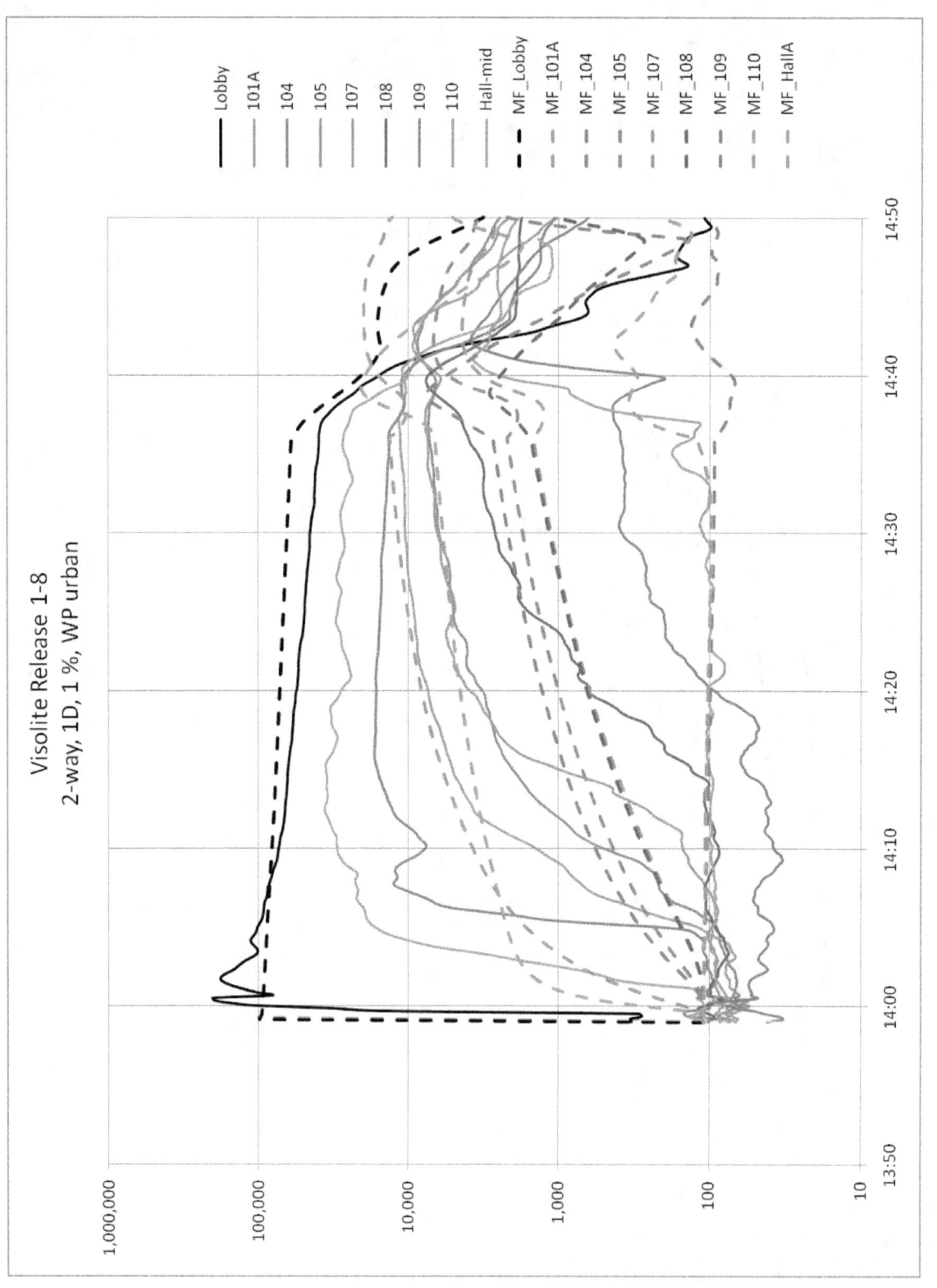

Figure 34 – Test 1-8

50

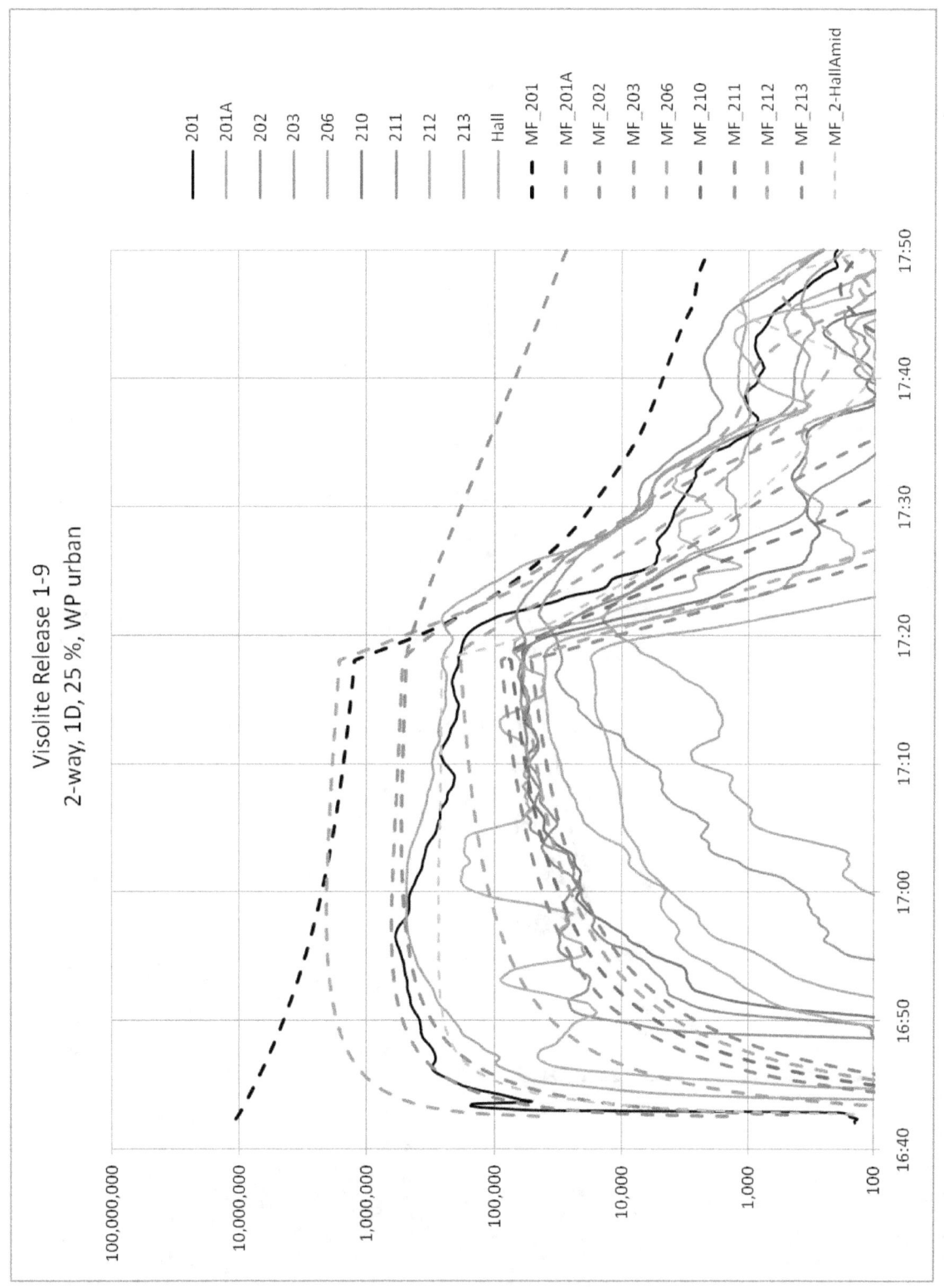

Figure 35 – Test 1-9

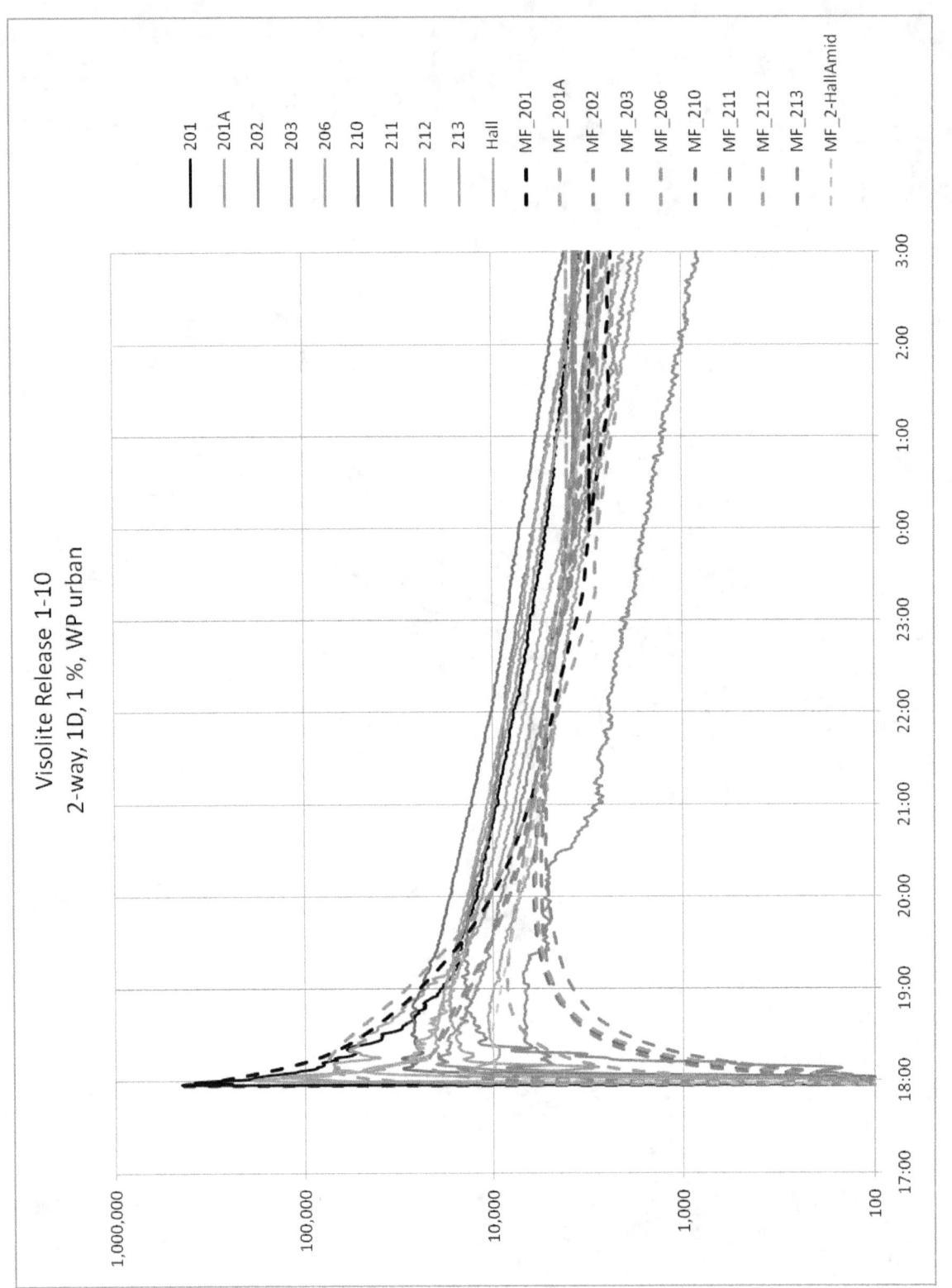

Visolite Release 1-10
2-way, 1D, 1 %, WP urban

Figure 36 – Test 1-10

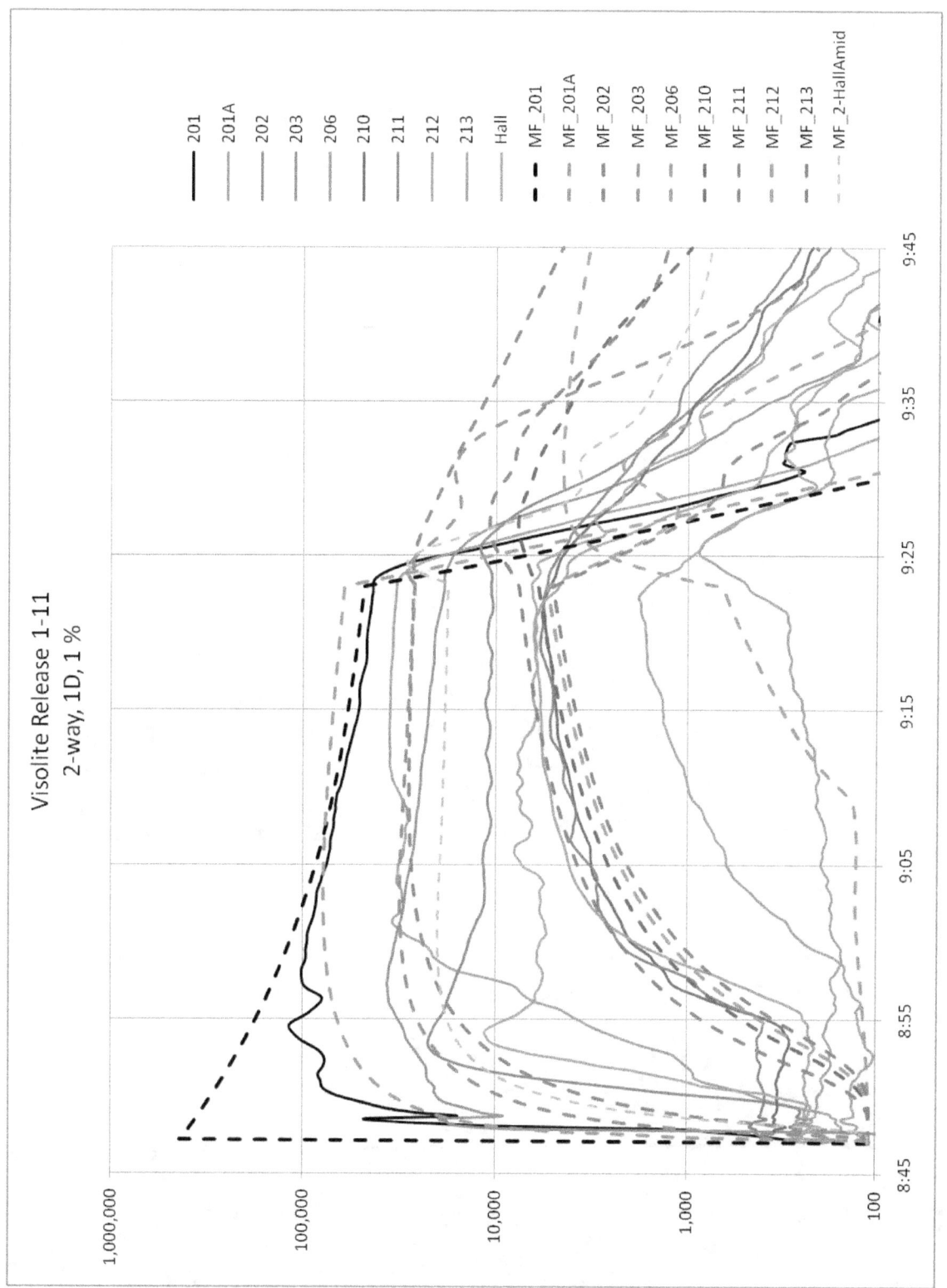

Figure 37 – Test 1-11

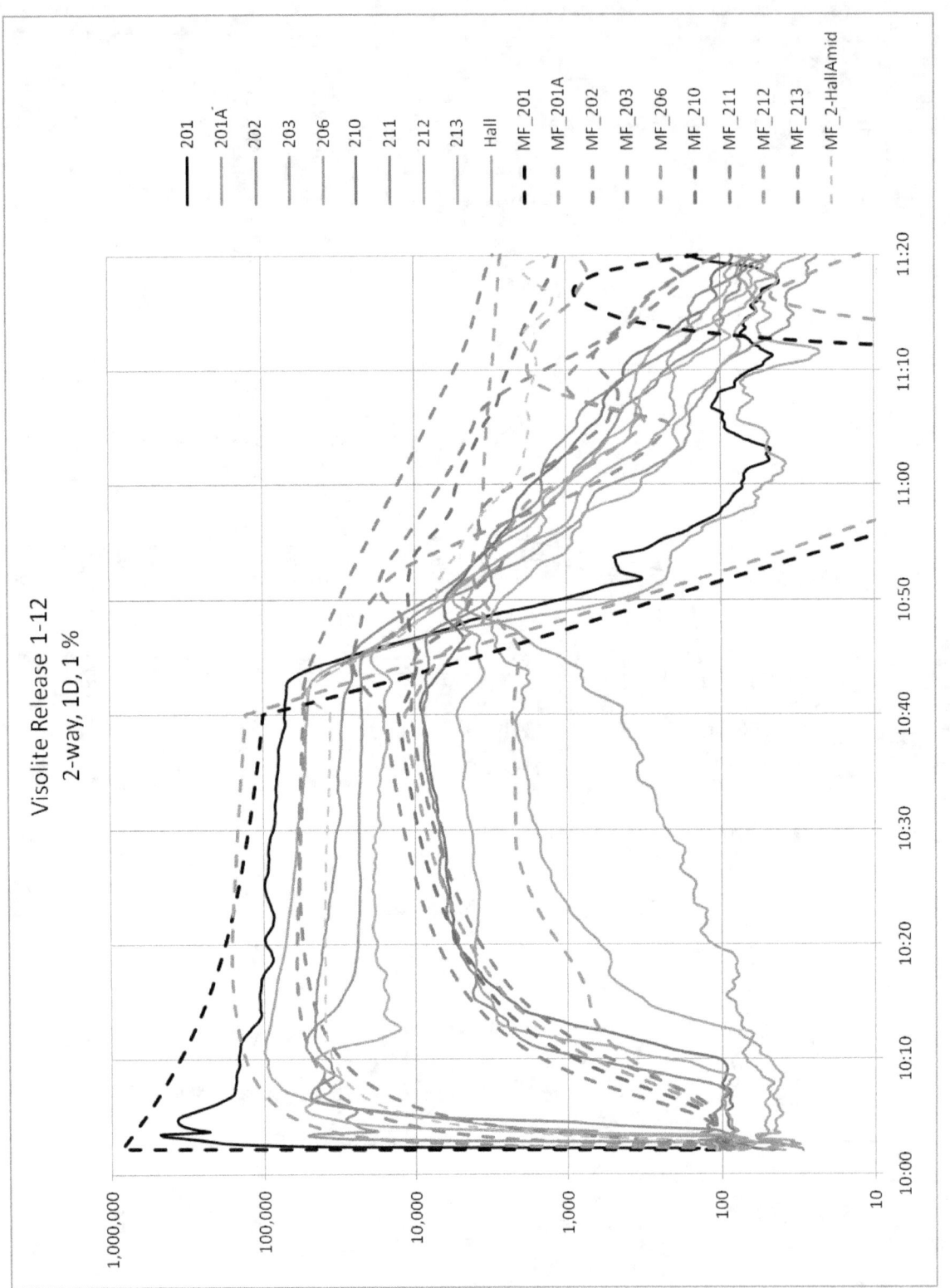

Figure 38 – Test 1-12

54

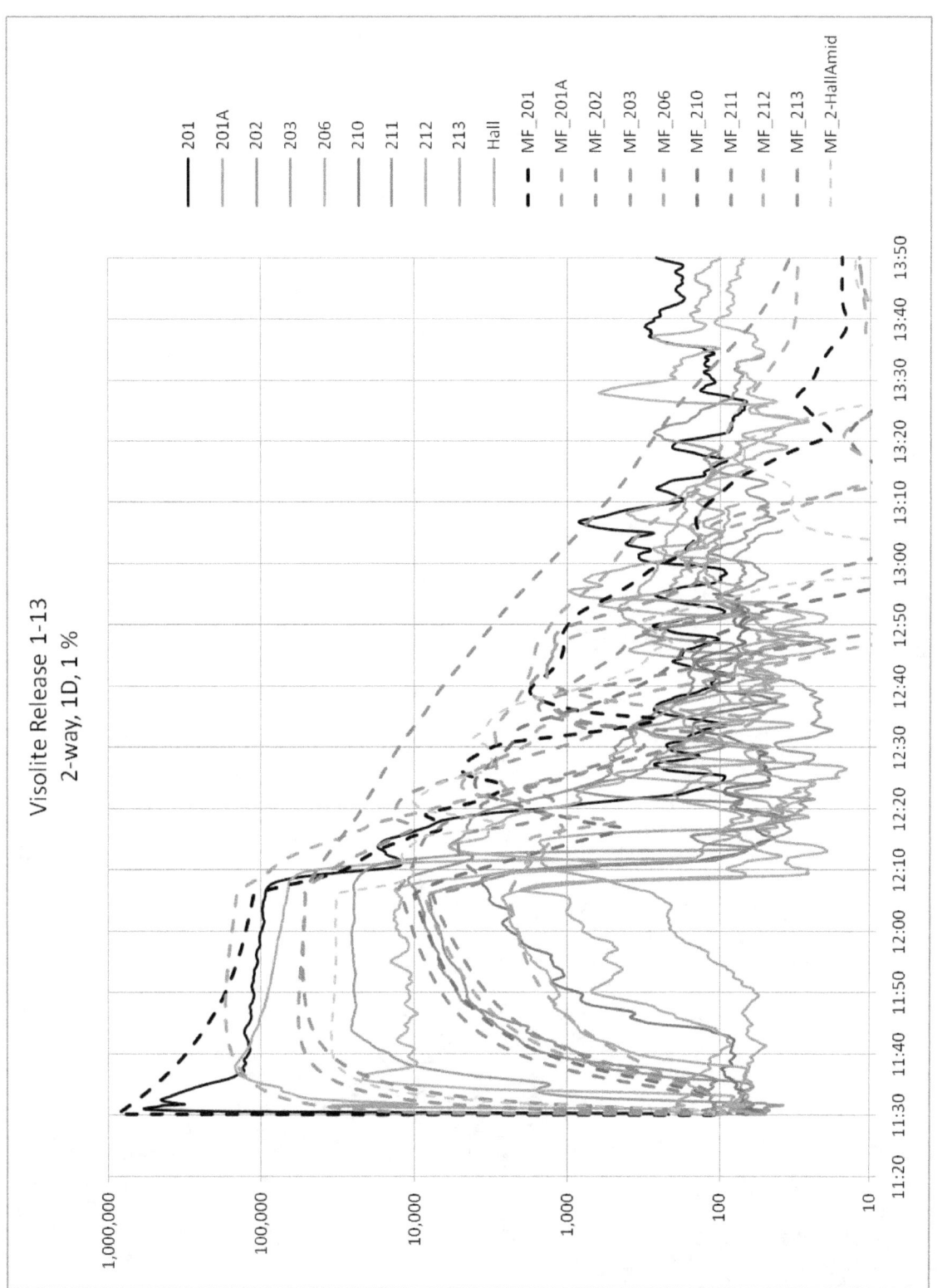

Figure 39 – Test 1-13

INL-2 Characterization Tests

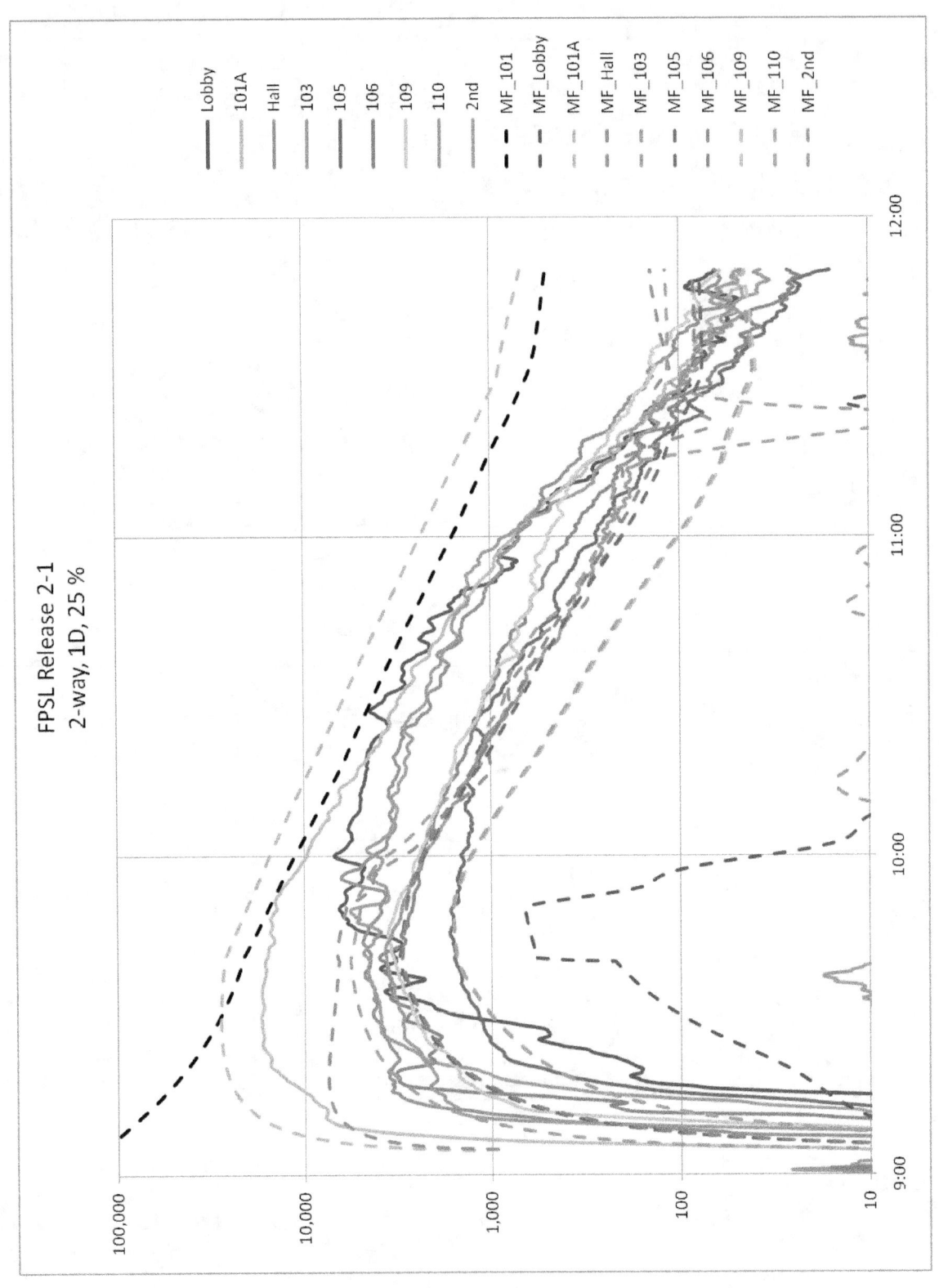

Figure 40 – Test 2-1 (measured data for source location 101 unavailable)

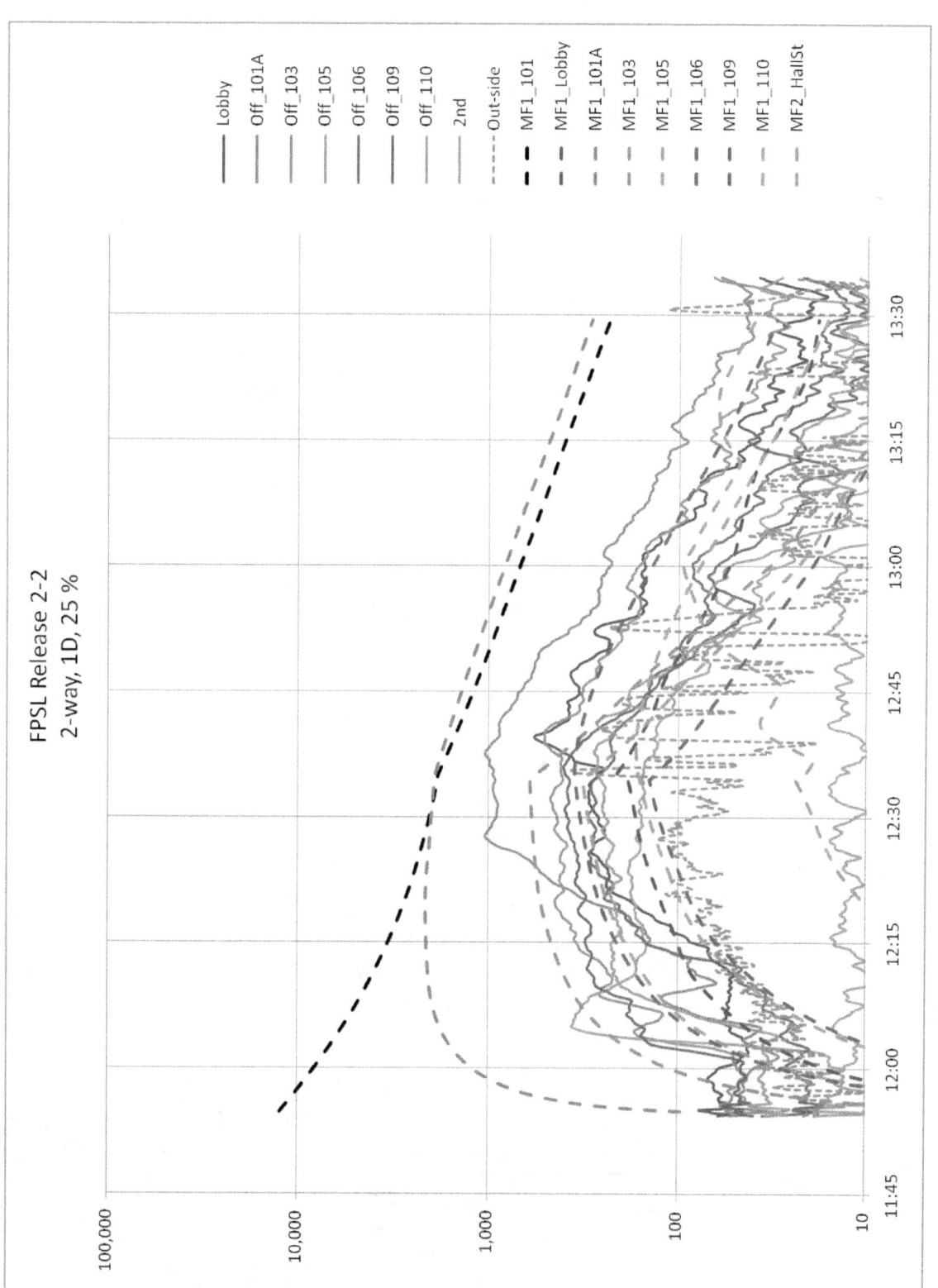

Figure 41 – Test 2-2 (measured data for source location 101 unavailable)

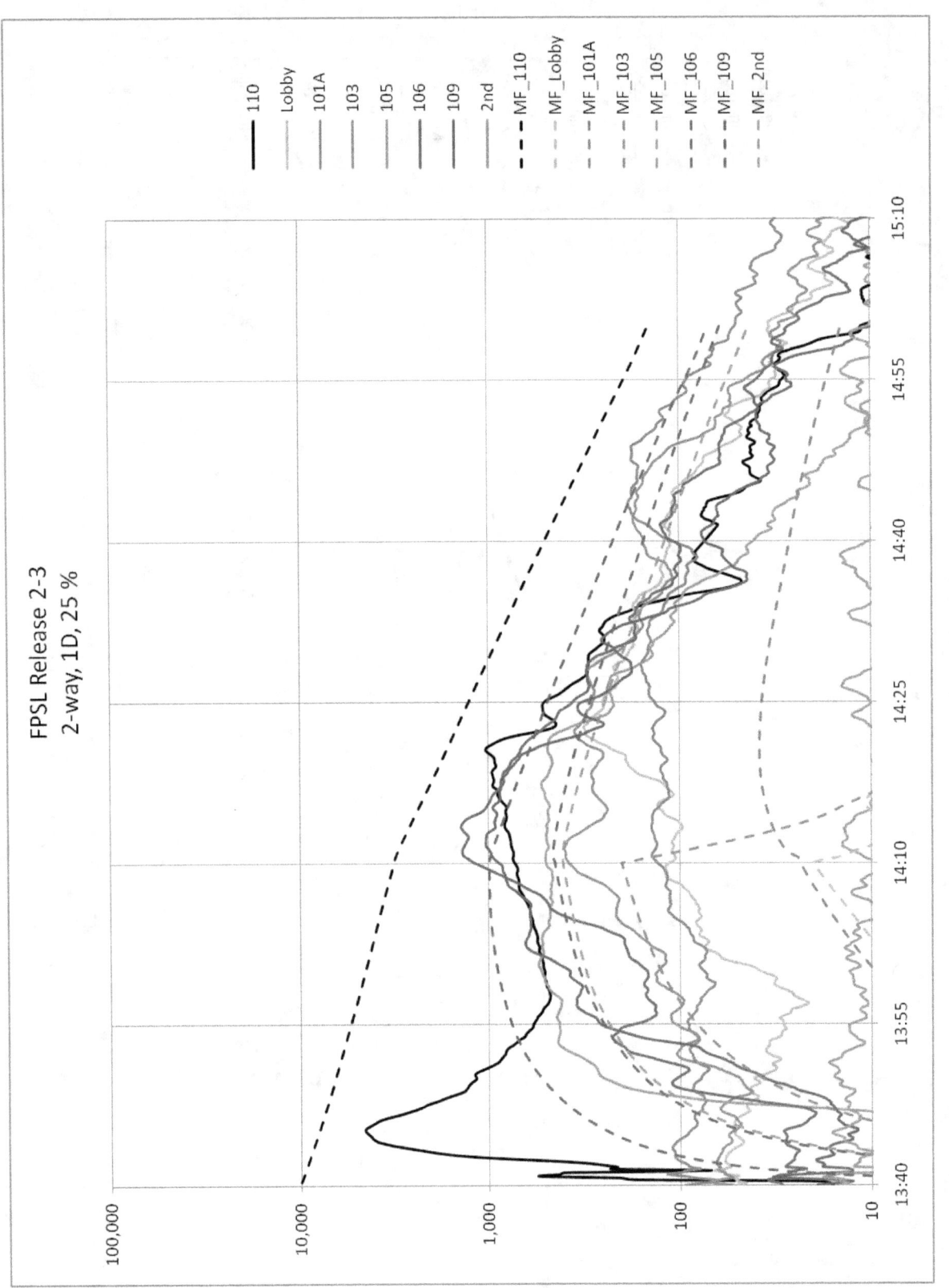

Figure 42 – Test 2-3

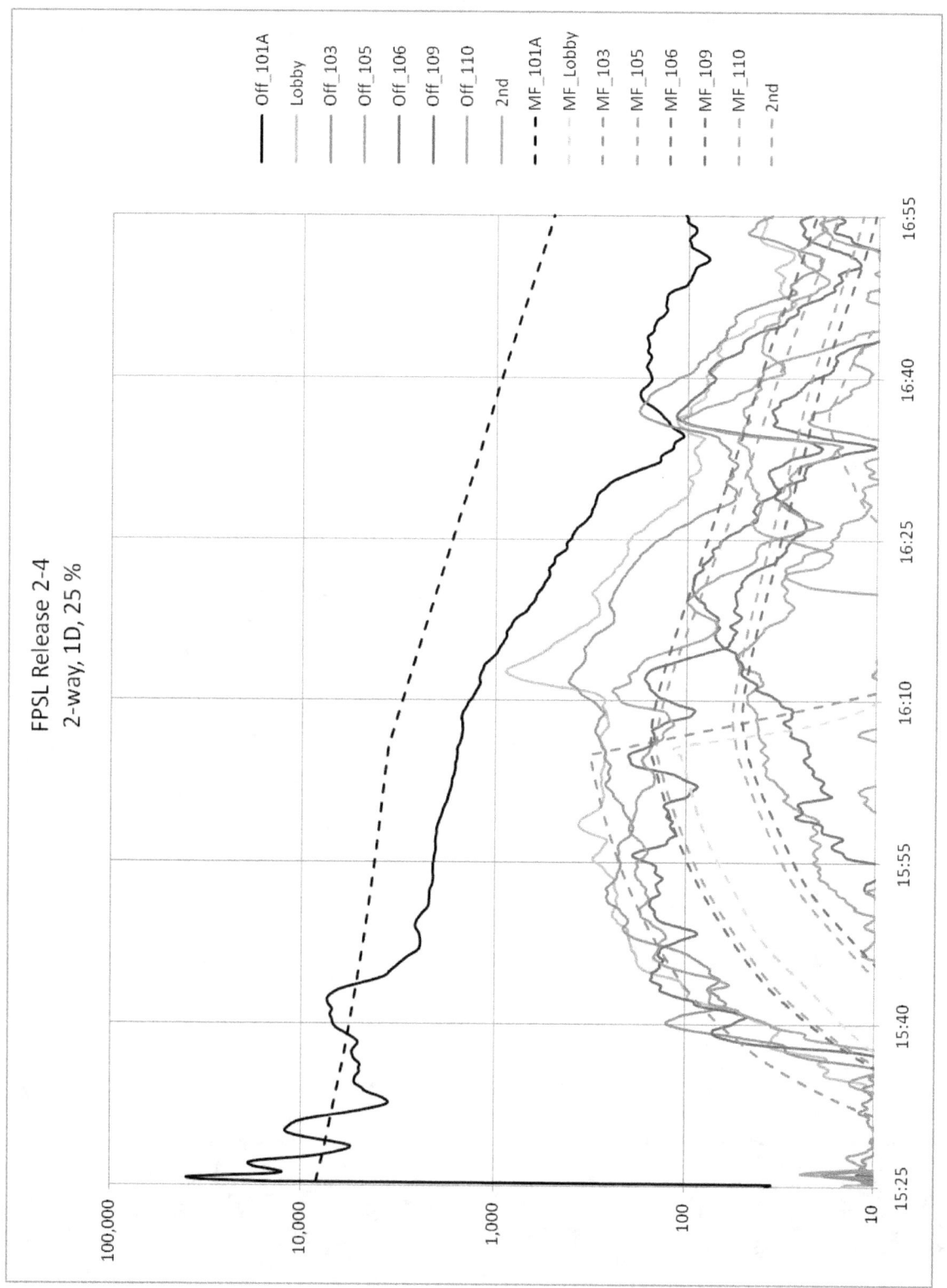

FPSL Release 2-4
2-way, 1D, 25 %

Off_101A	MF_101A
Lobby	MF_Lobby
Off_103	MF_103
Off_105	MF_105
Off_106	MF_106
Off_109	MF_109
Off_110	MF_110
2nd	2nd

Figure 43 – Test 2-4

59

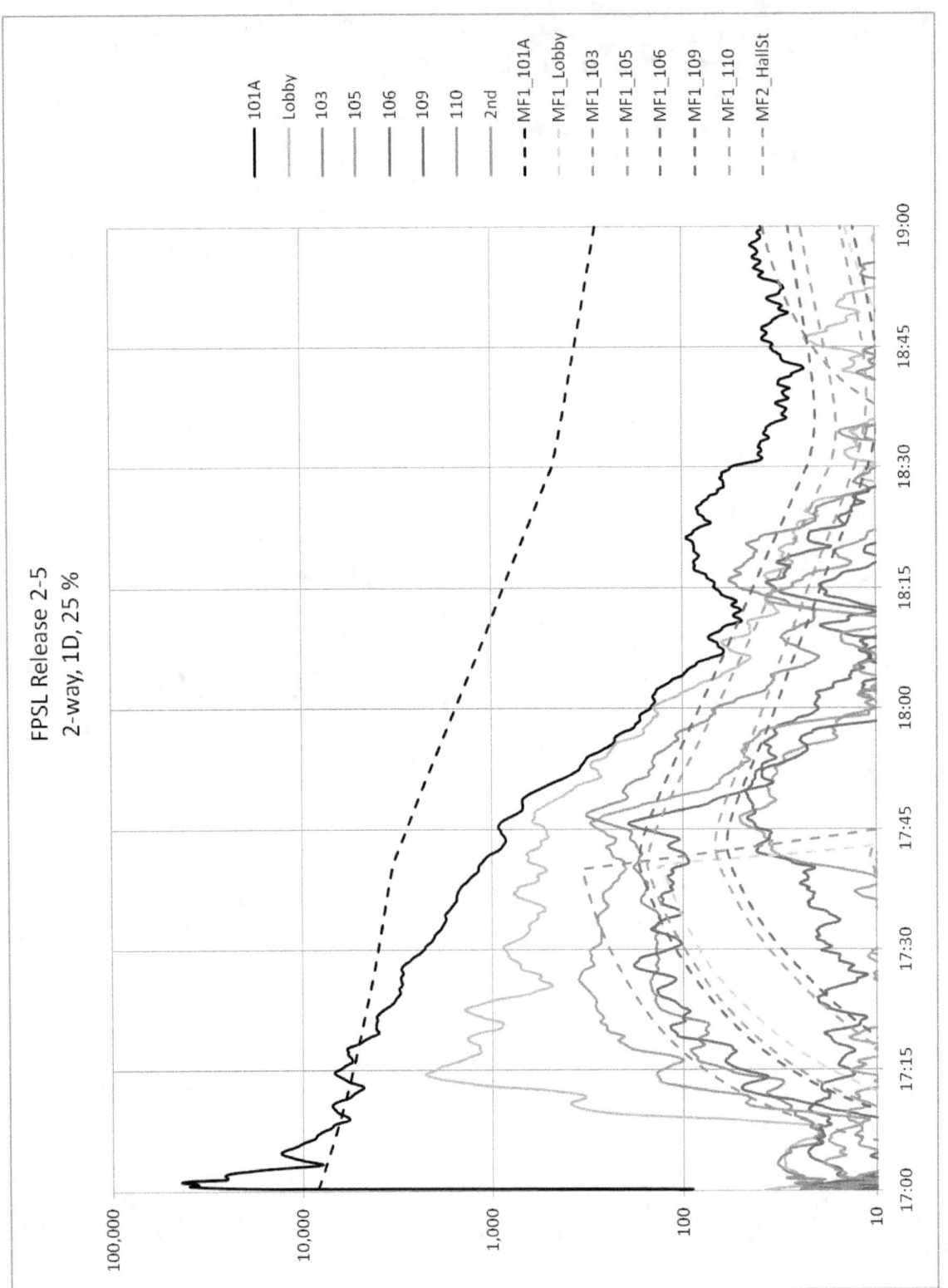

Figure 44 – Test 2-5

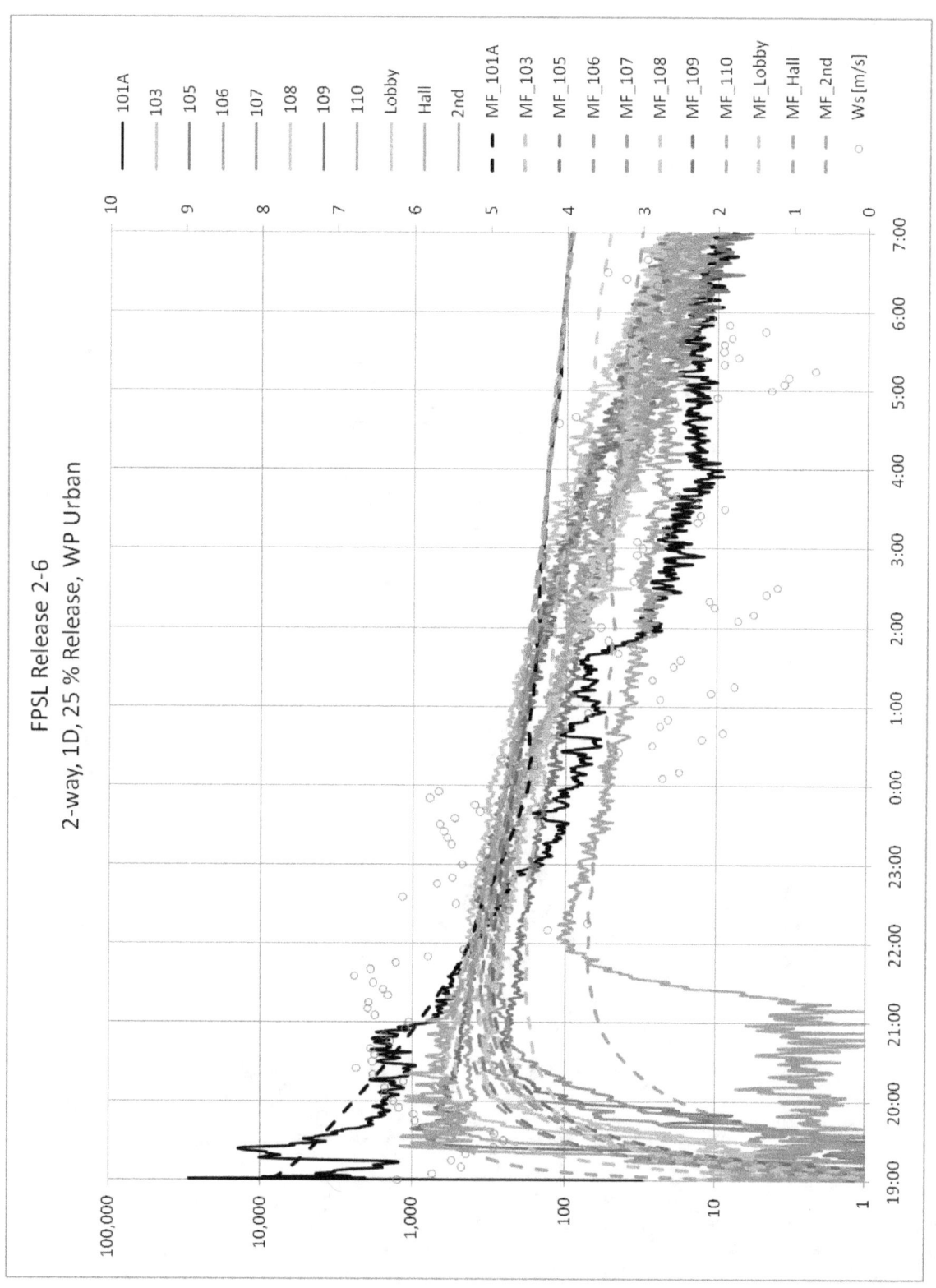

Figure 45 – Test 2-6 (urban wind pressure terrain effects)

61

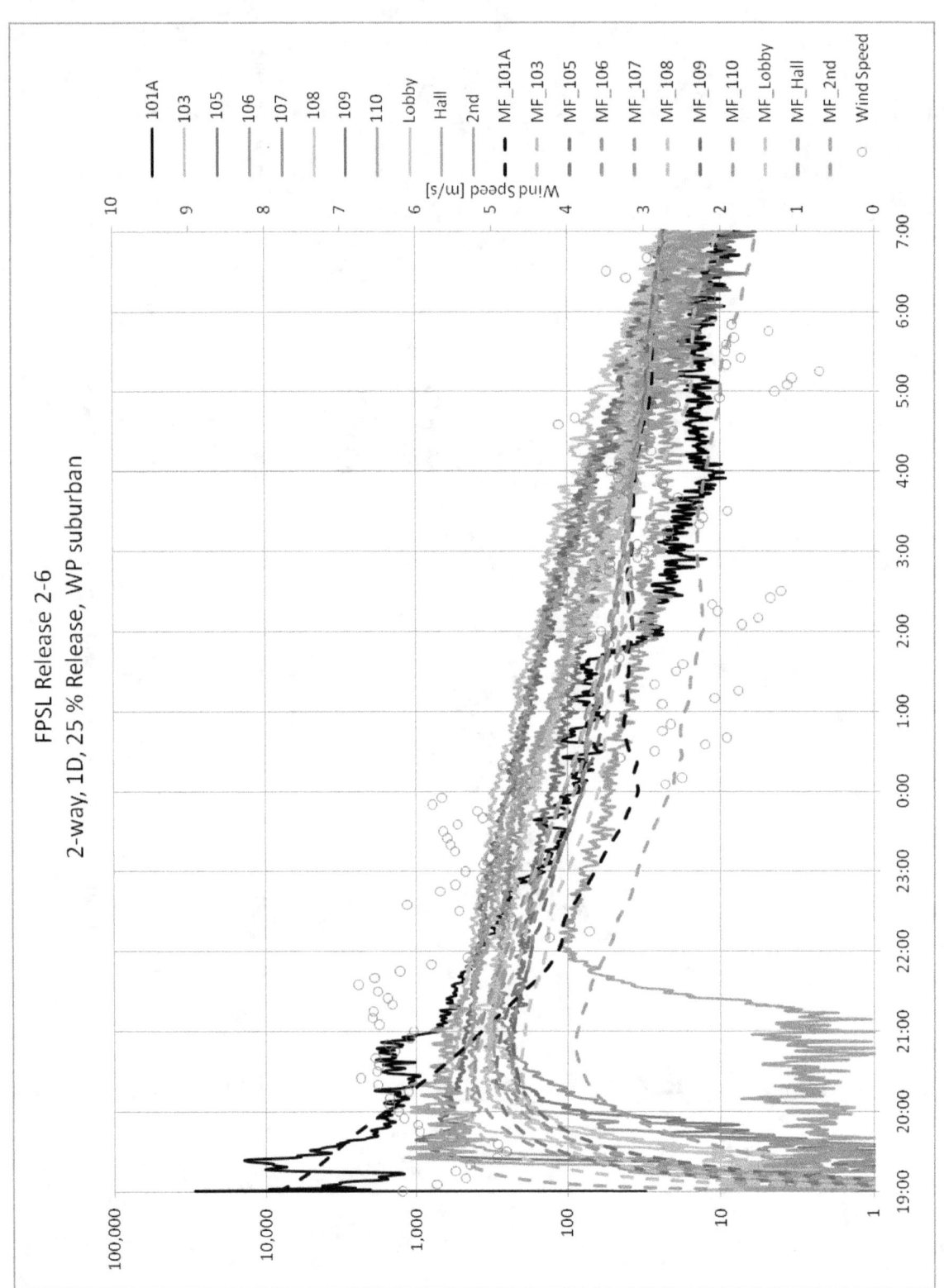

Figure 46 – Test 2-6 (suburban wind pressure terrain effects)

62

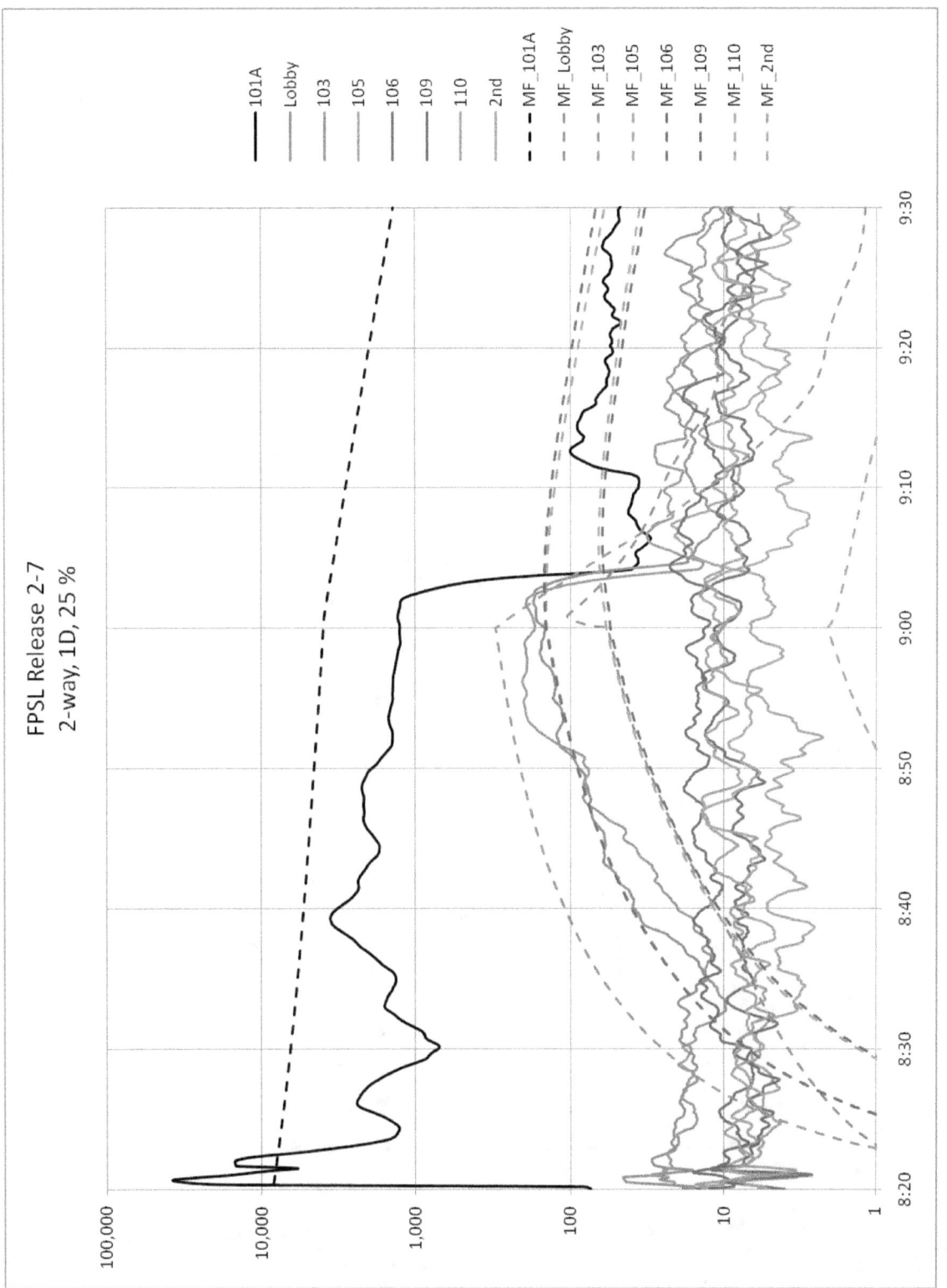

FPSL Release 2-7
2-way, 1D, 25 %

Figure 47 – Test 2-7

63

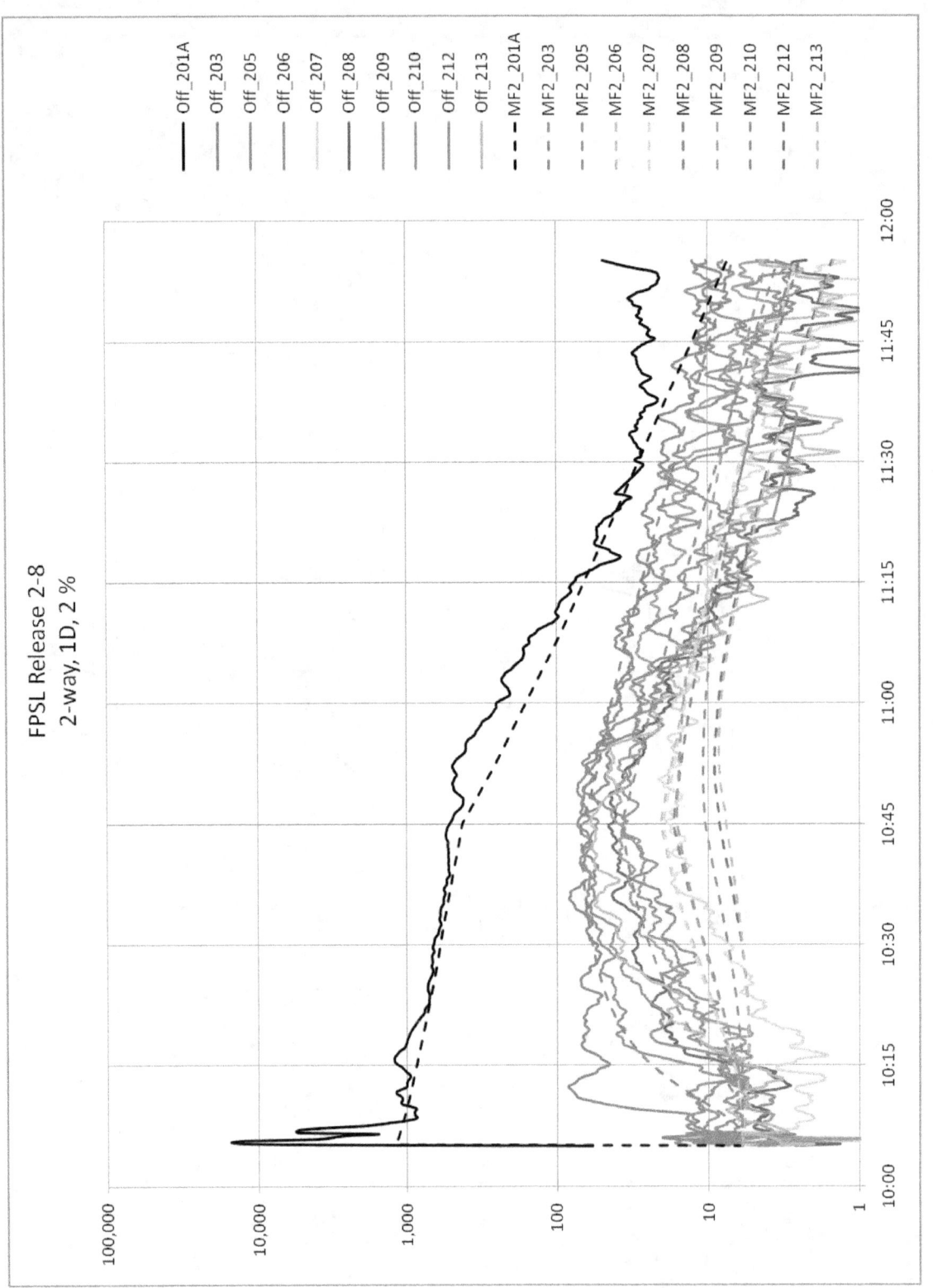

Figure 48 – Test 2-8

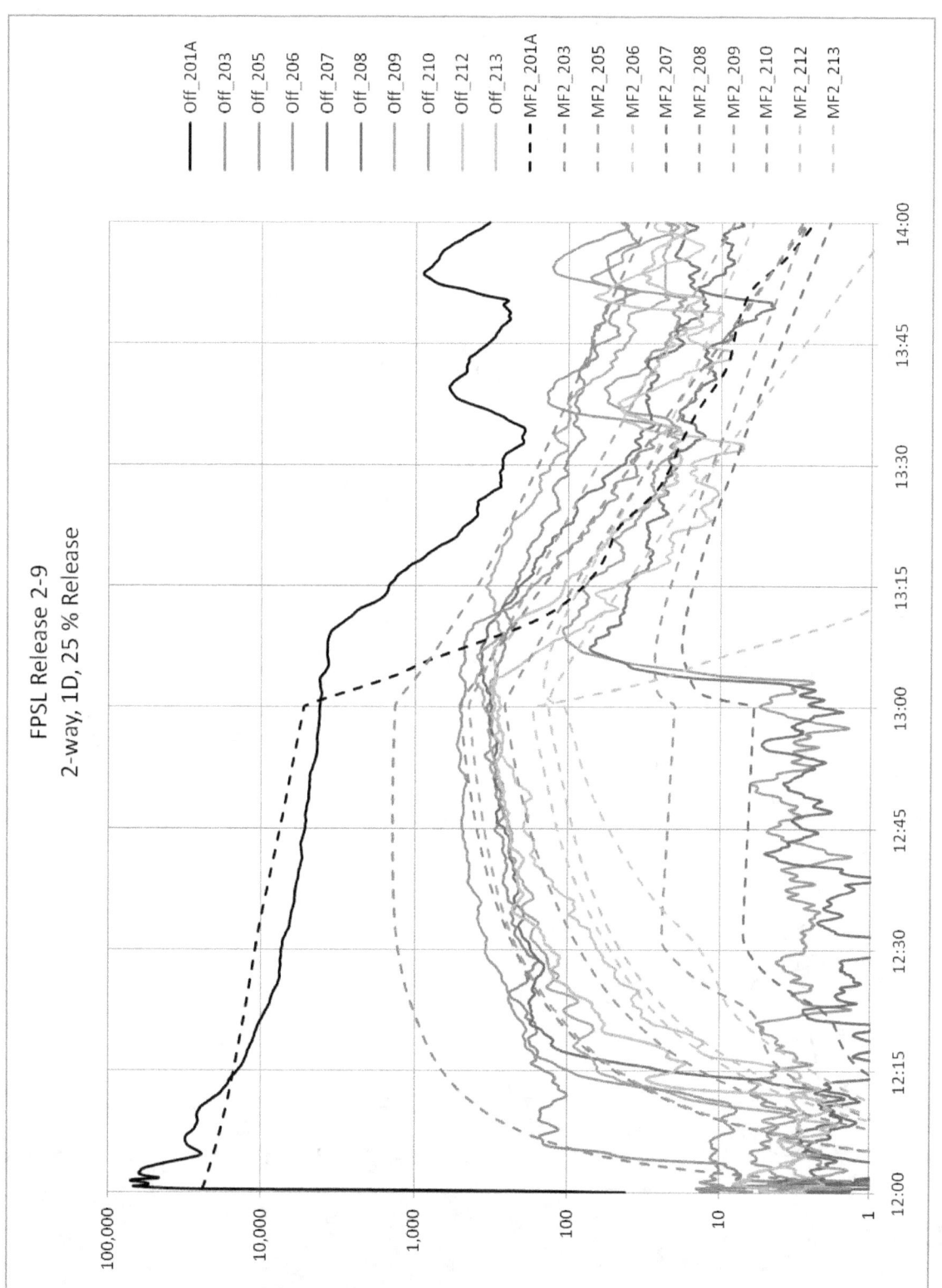

Figure 49 – Test 2-9

65

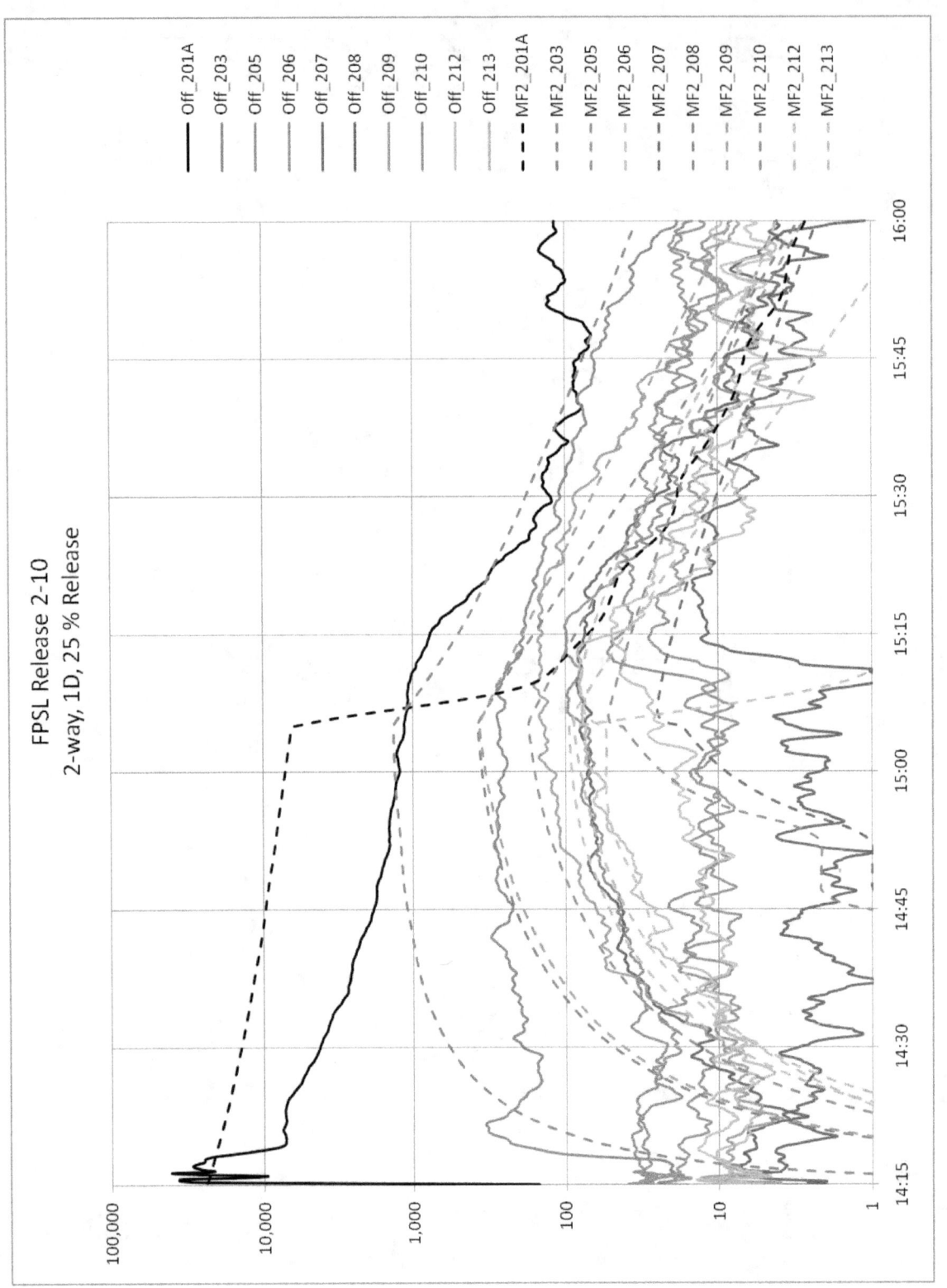

FPSL Release 2-10
2-way, 1D, 25 % Release

Legend:
Off_201A, Off_203, Off_205, Off_206, Off_207, Off_208, Off_209, Off_210, Off_212, Off_213, MF2_201A, MF2_203, MF2_205, MF2_206, MF2_207, MF2_208, MF2_209, MF2_210, MF2_212, MF2_213

Figure 50 – Test 2-10

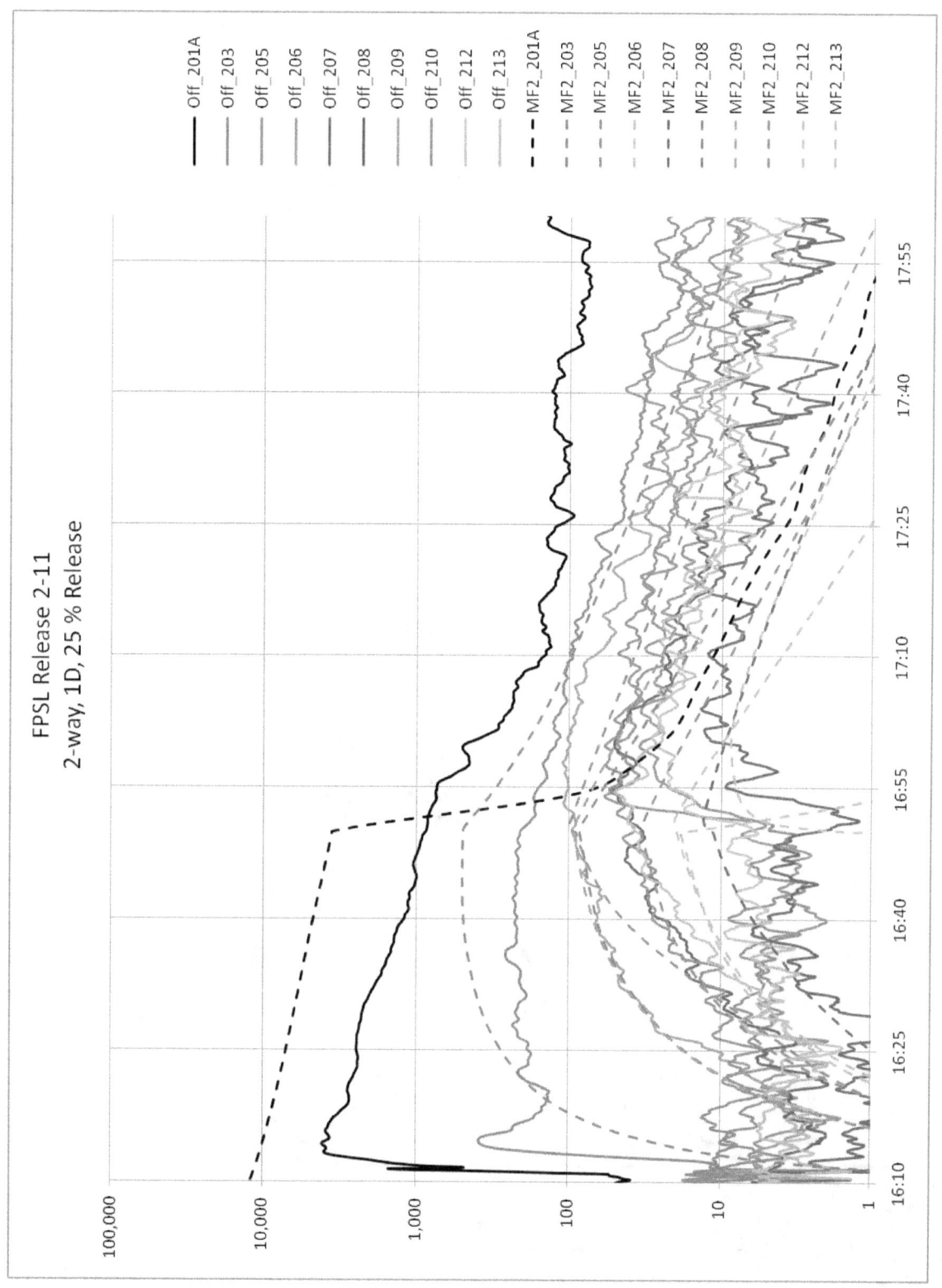

Figure 51 – Test 2-11

67

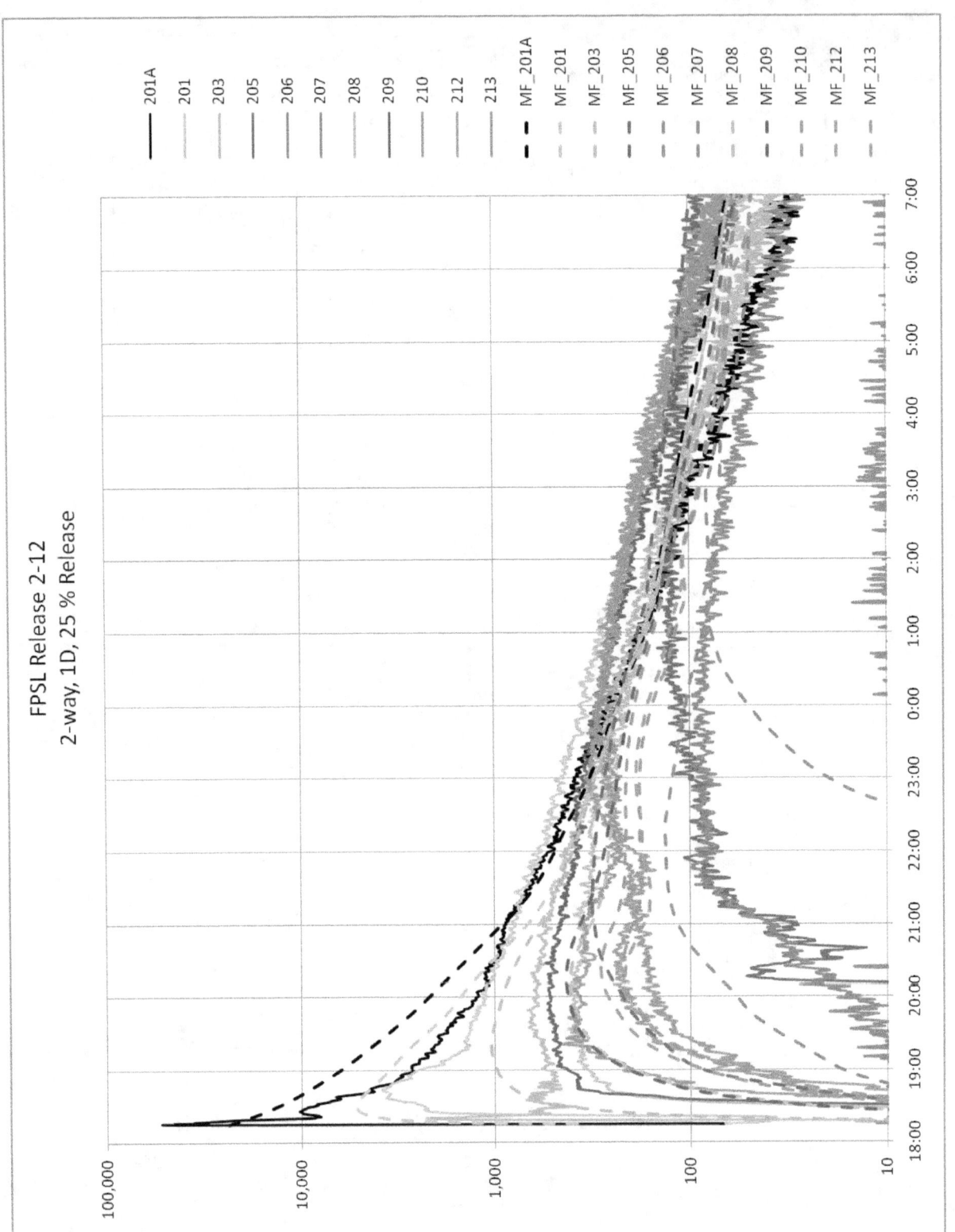

Figure 52 – Test 2-12

Appendix B – Measured vs. Predicted Average Concentration

This appendix provides bar charts displaying measured versus predicted average particle concentrations of all building characterization tests. Simulations were performed employing the 1D convection-diffusion, two-way openings and reduced release amounts as provided in the chart headings. These charts provide a quick visual means to determine order-of-magnitude comparisons between measured and predicted values.

INL-1 Characterization Tests

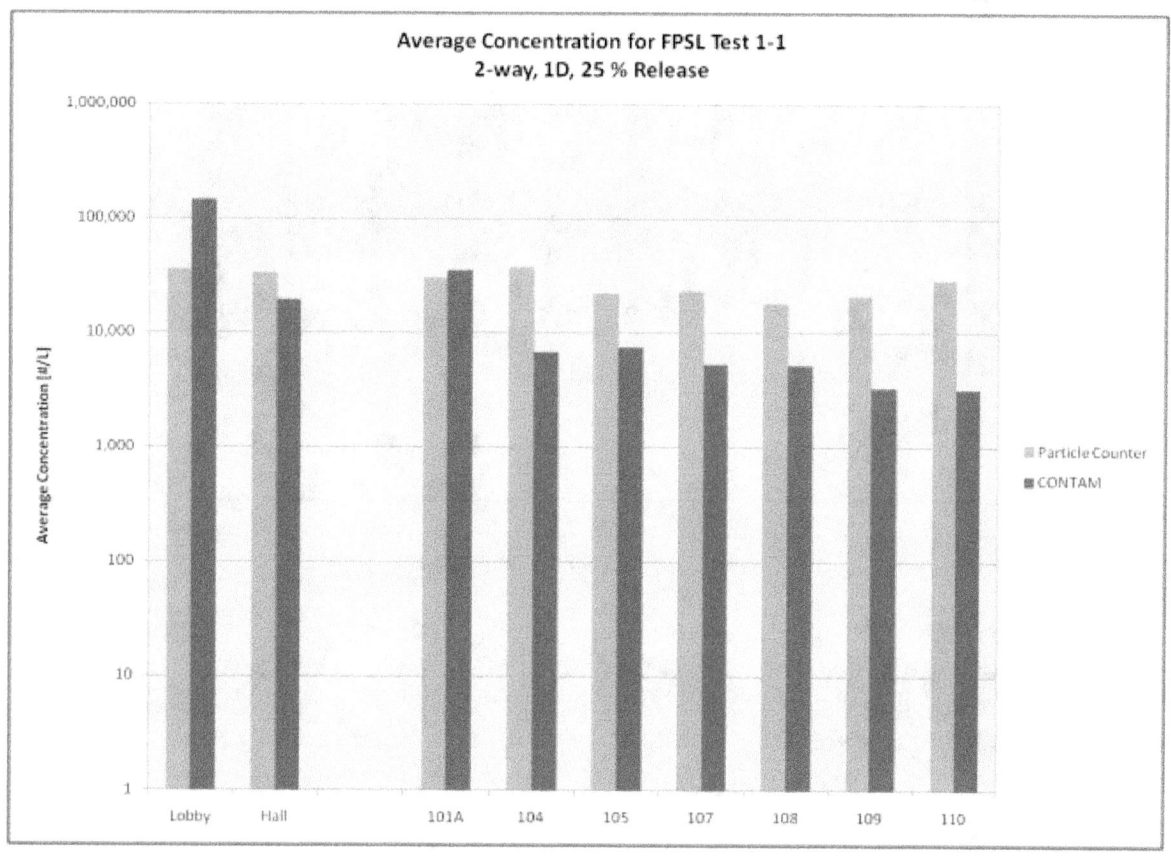

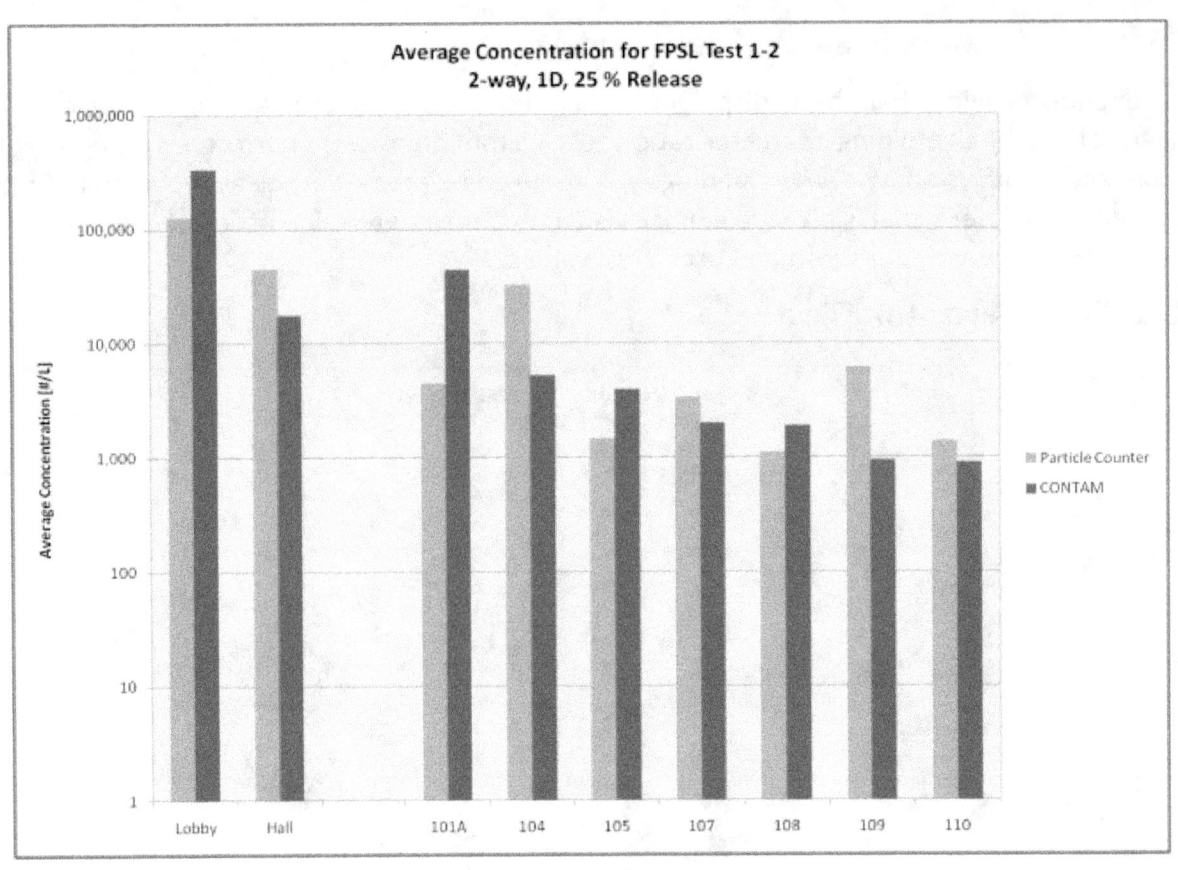

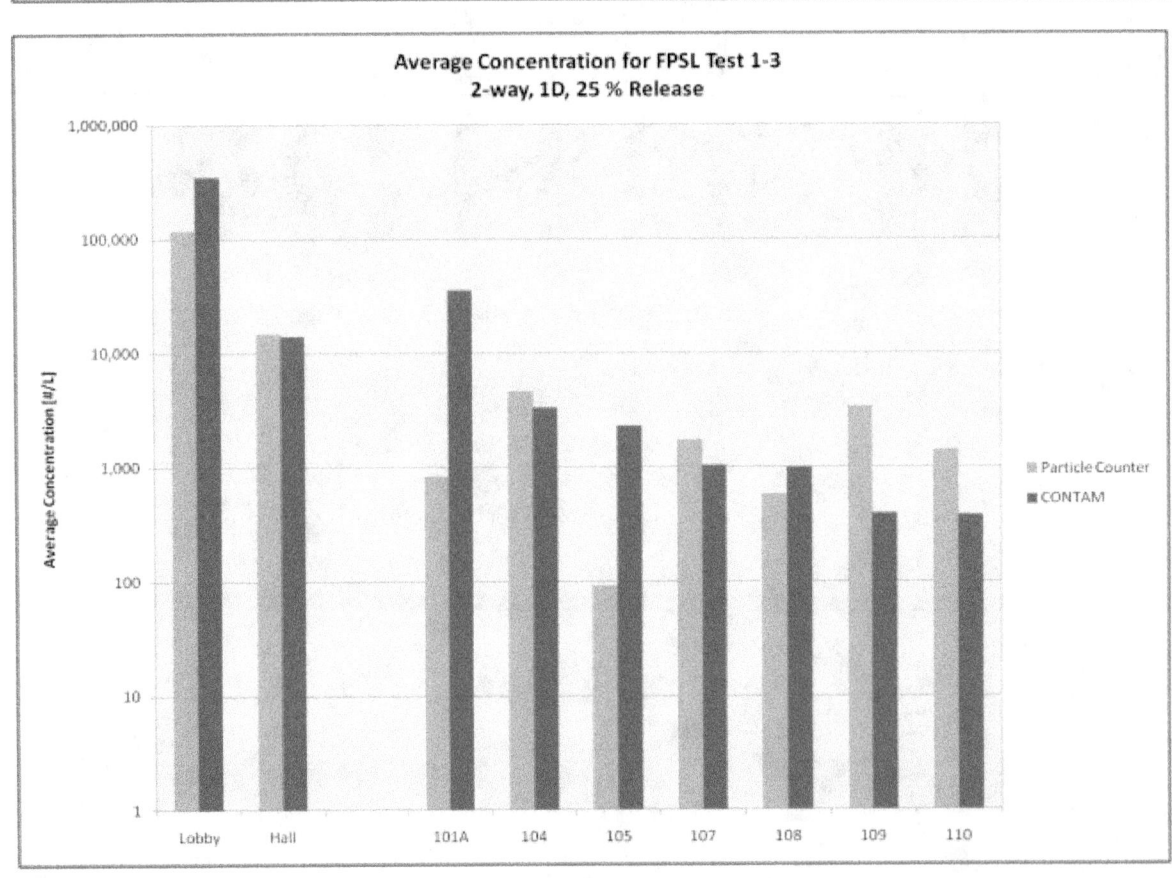

70

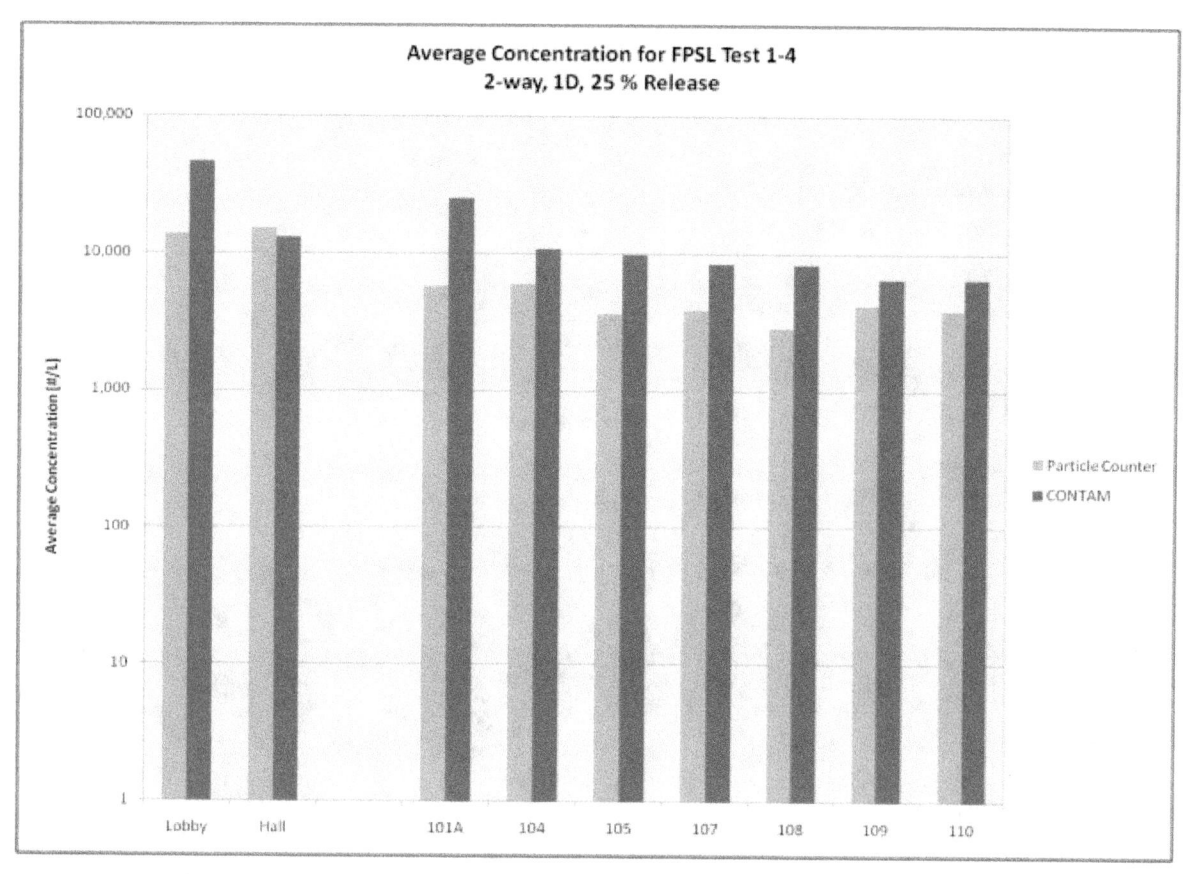

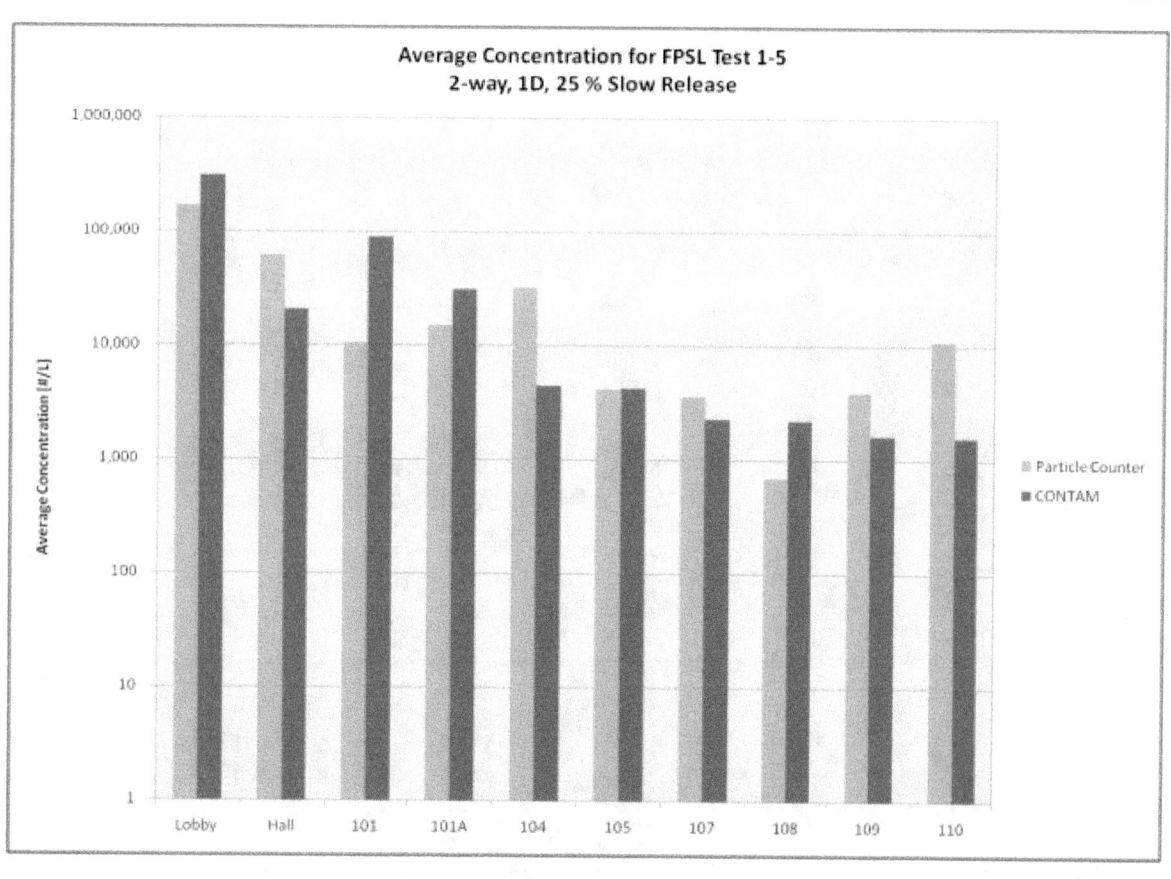

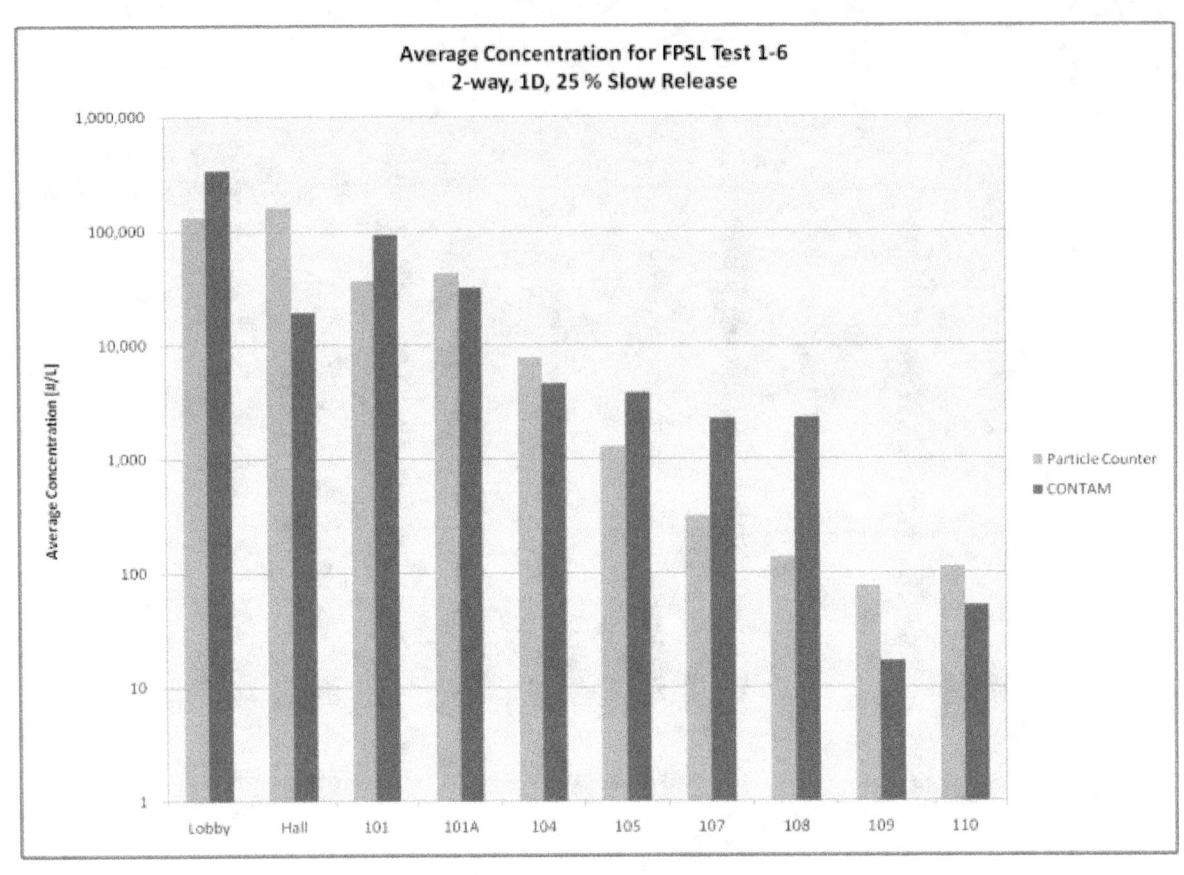

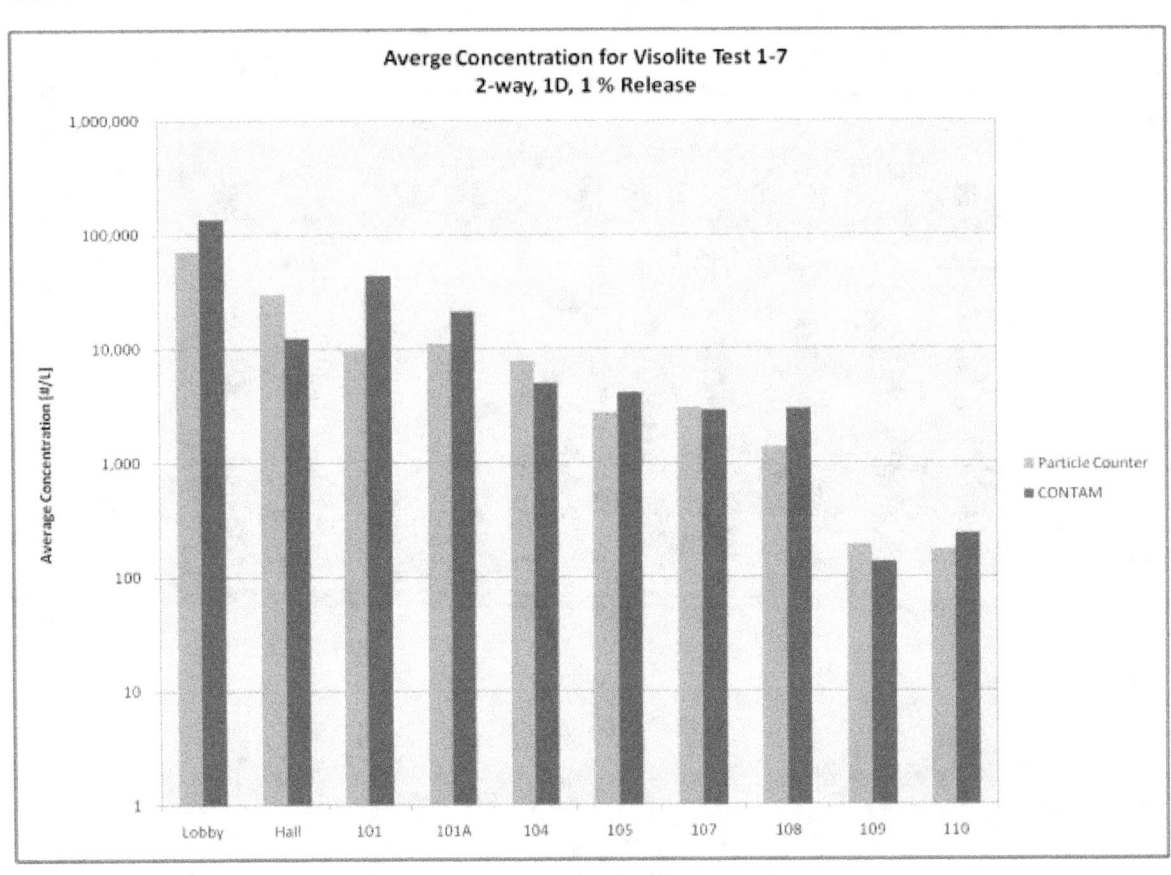

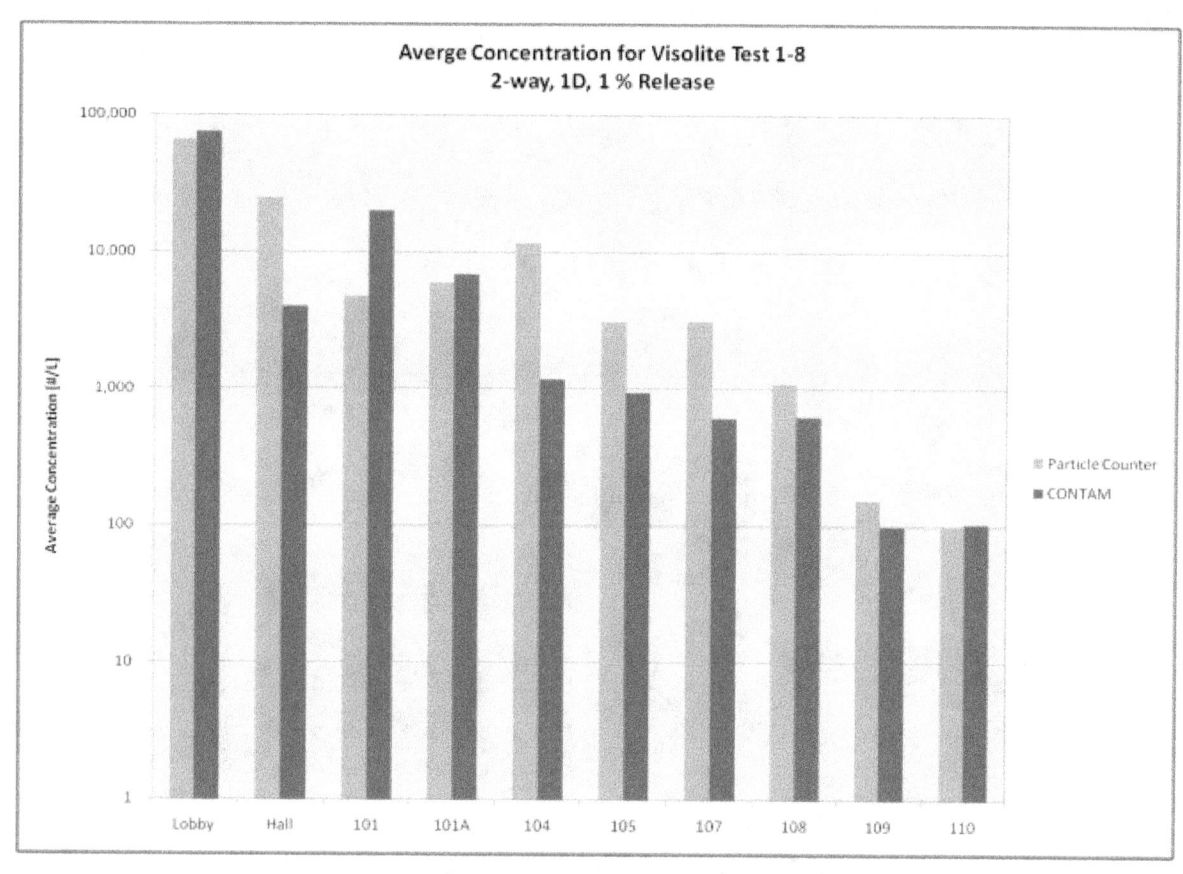

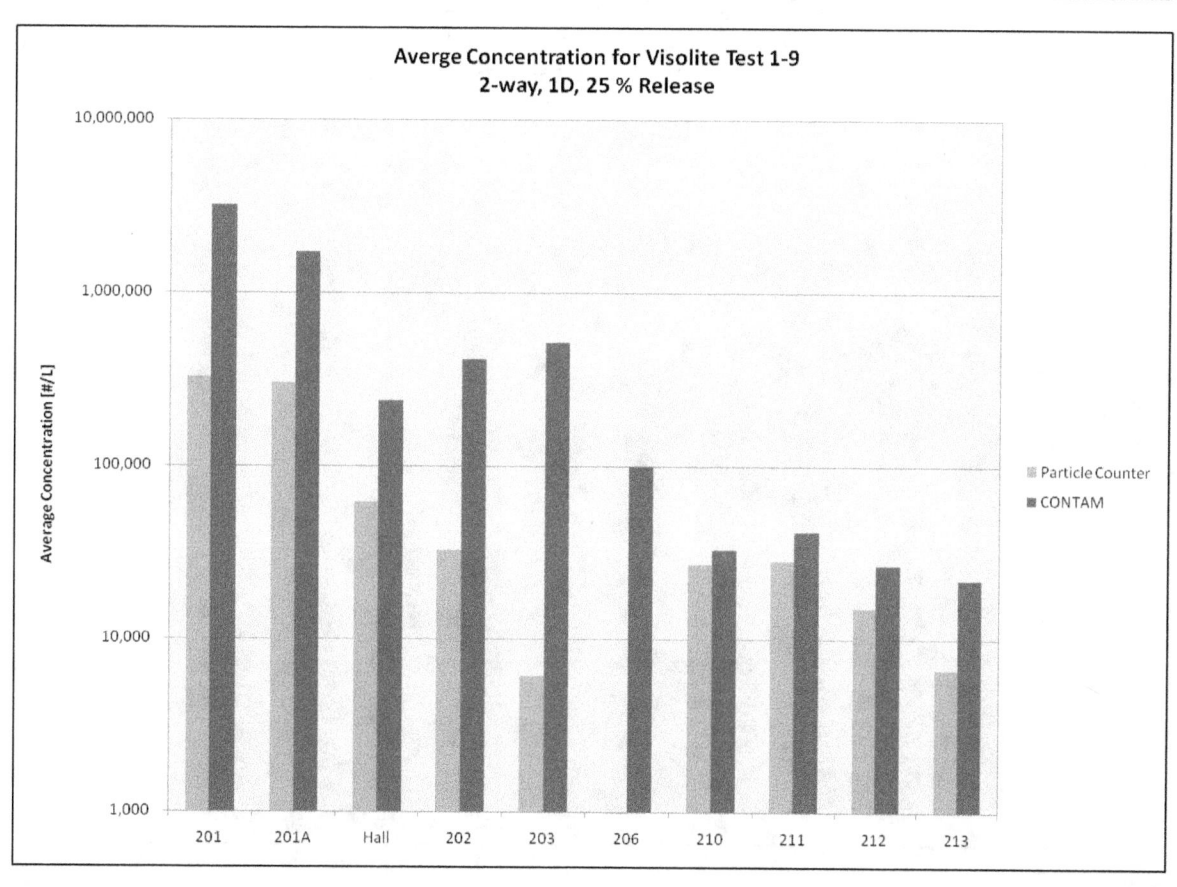

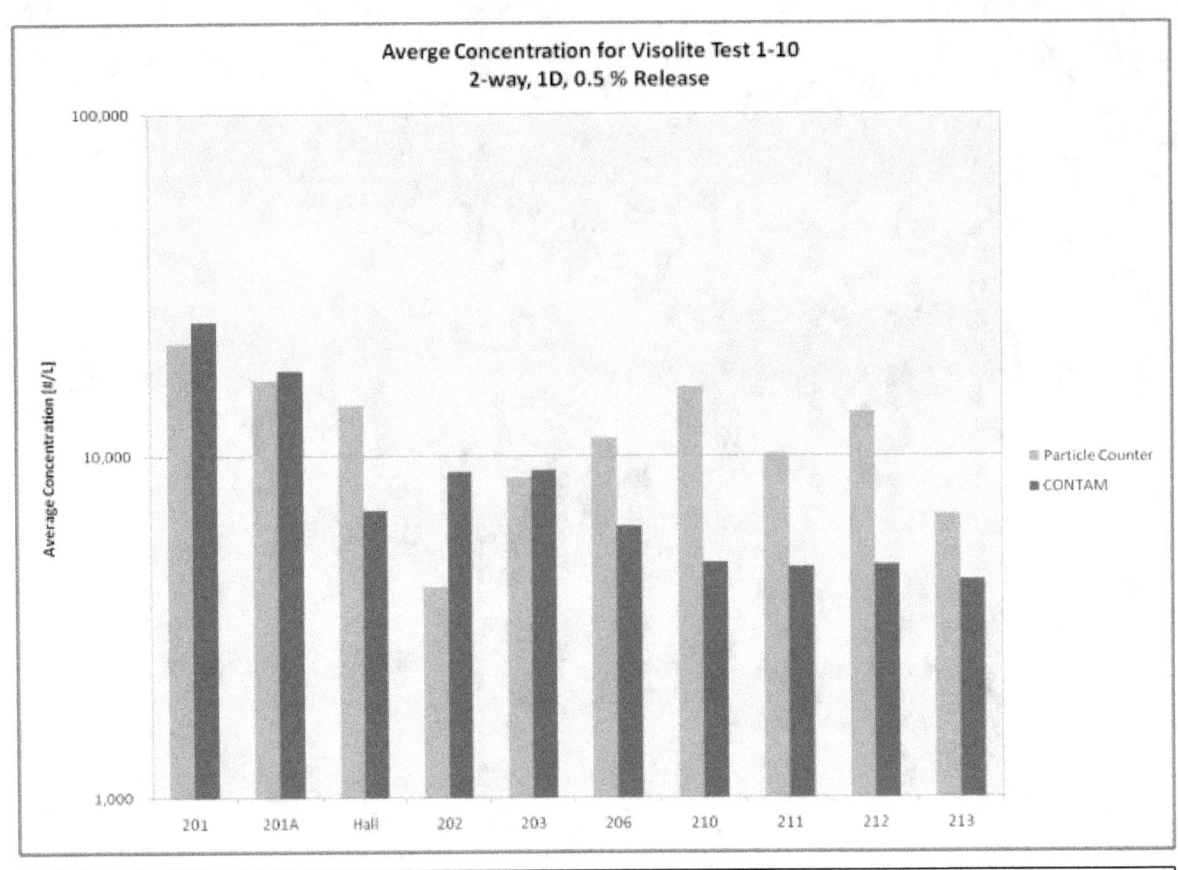

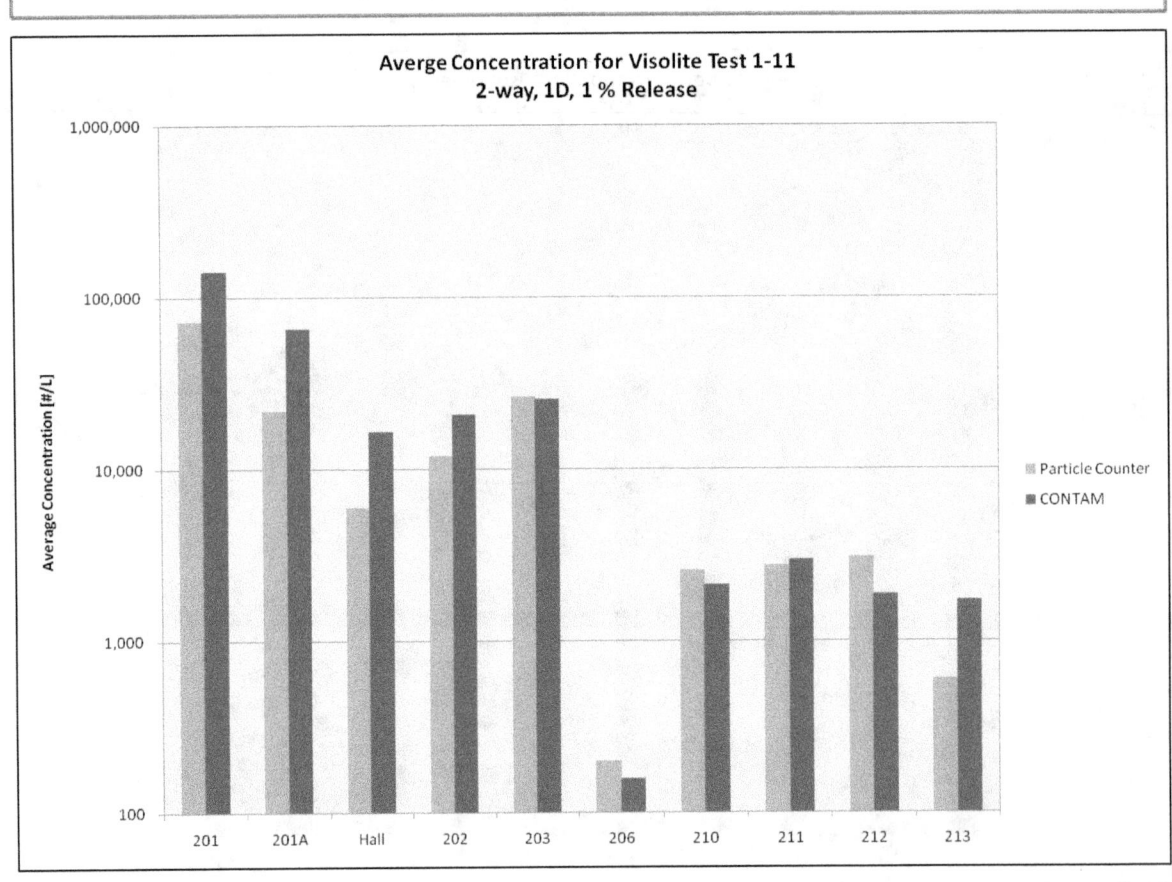

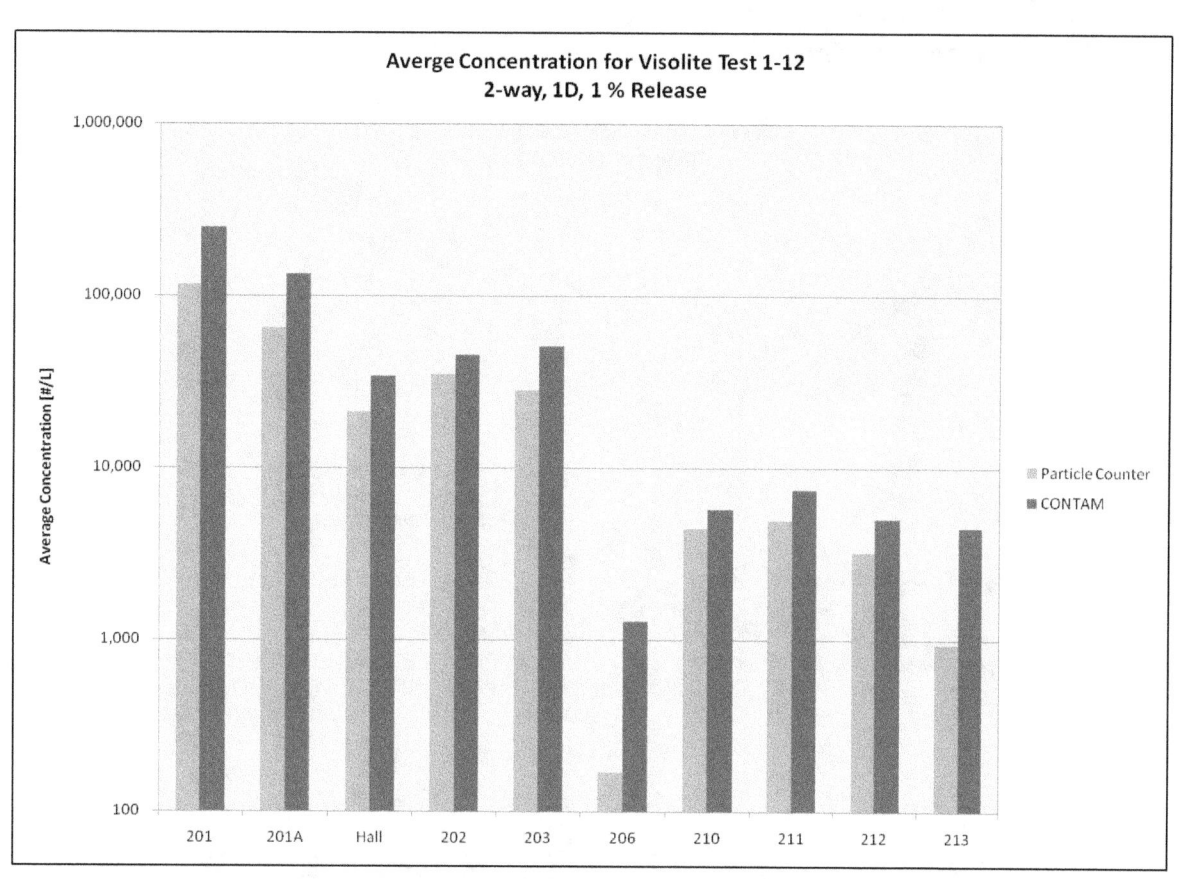

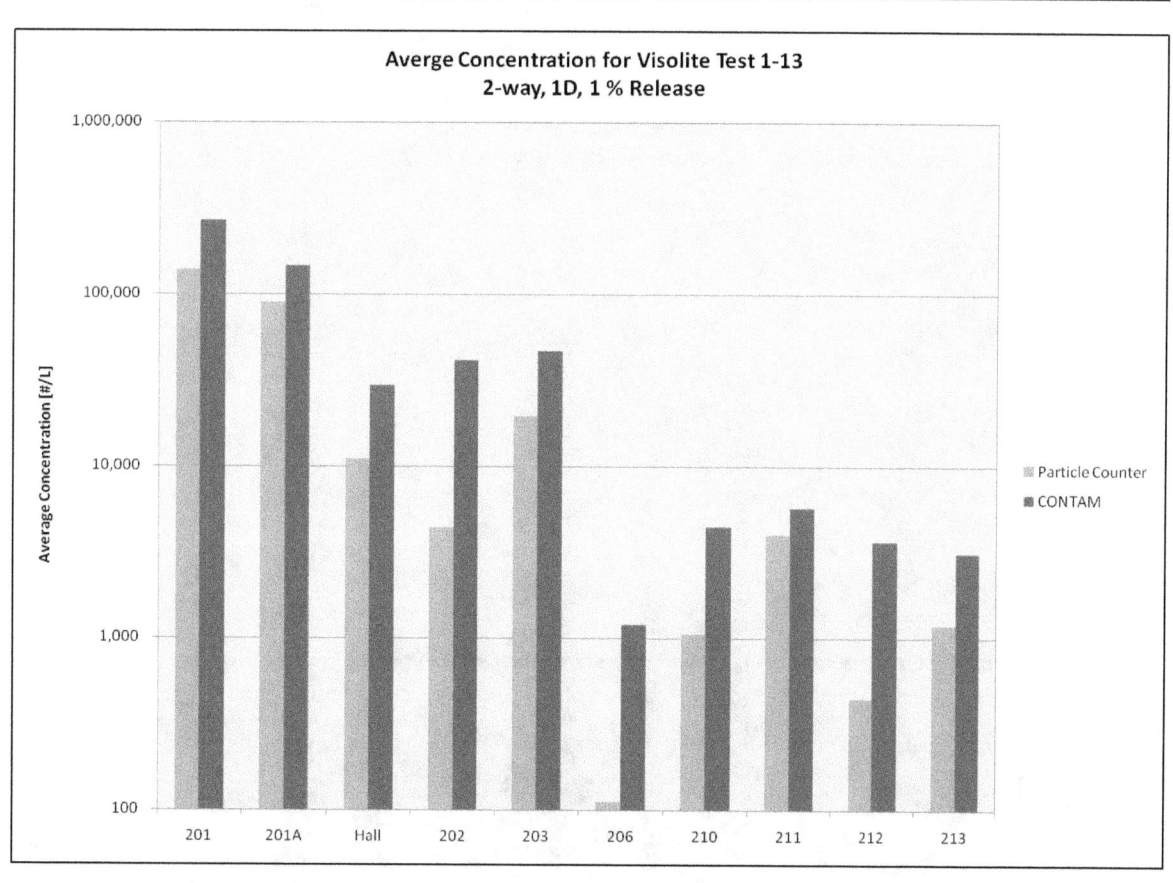

INL-2 Characterization Tests

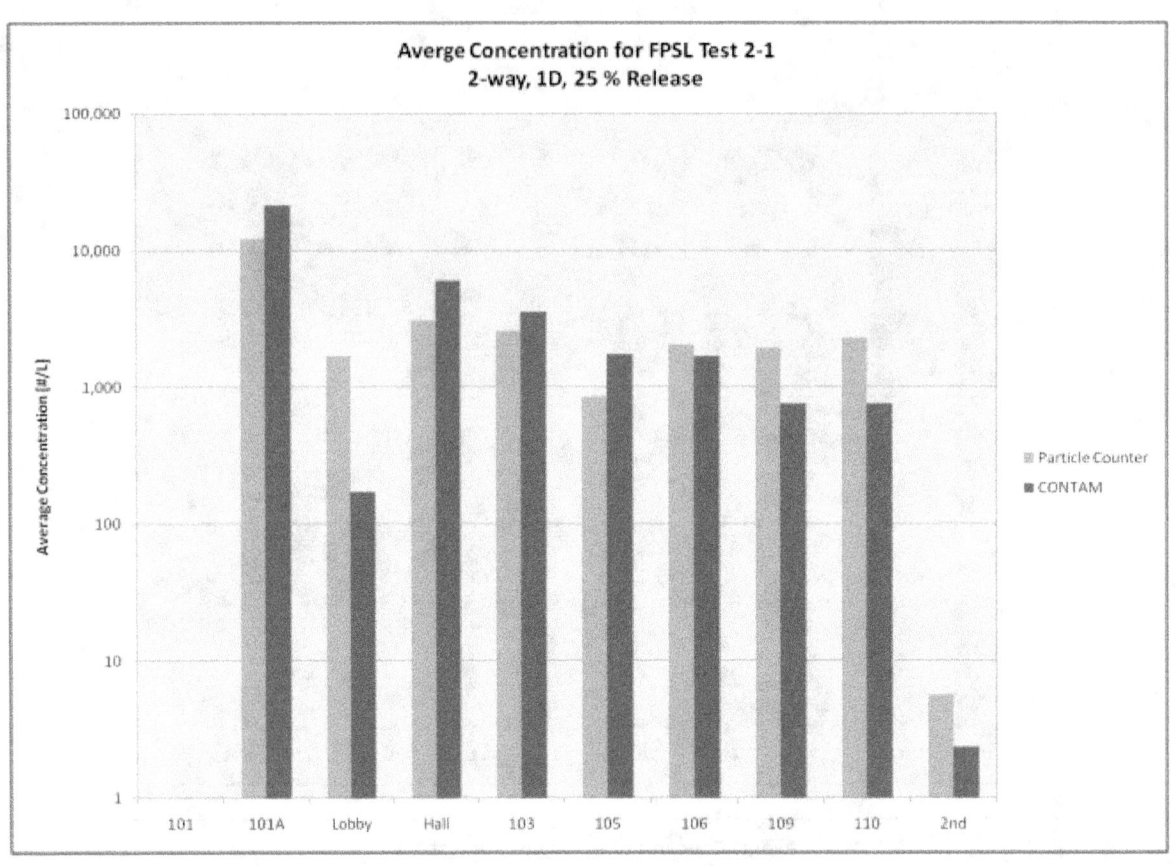

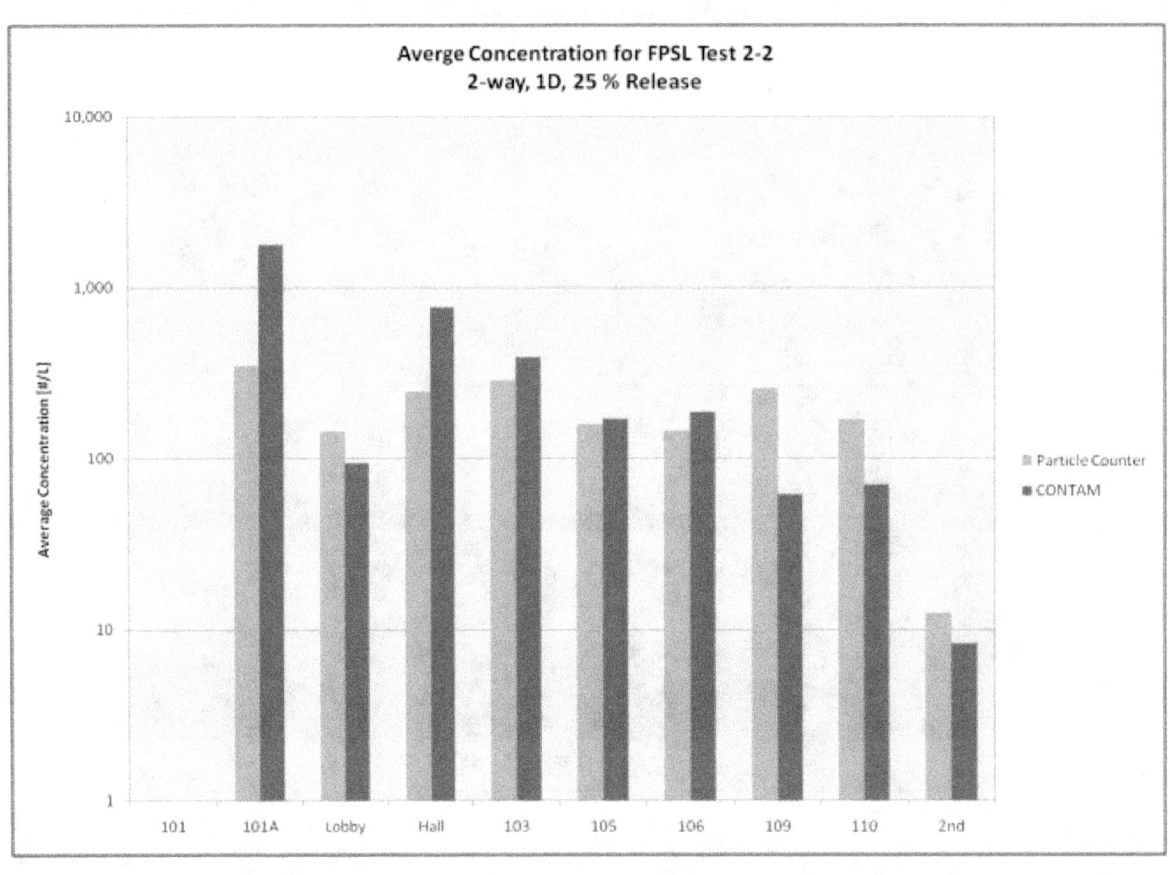

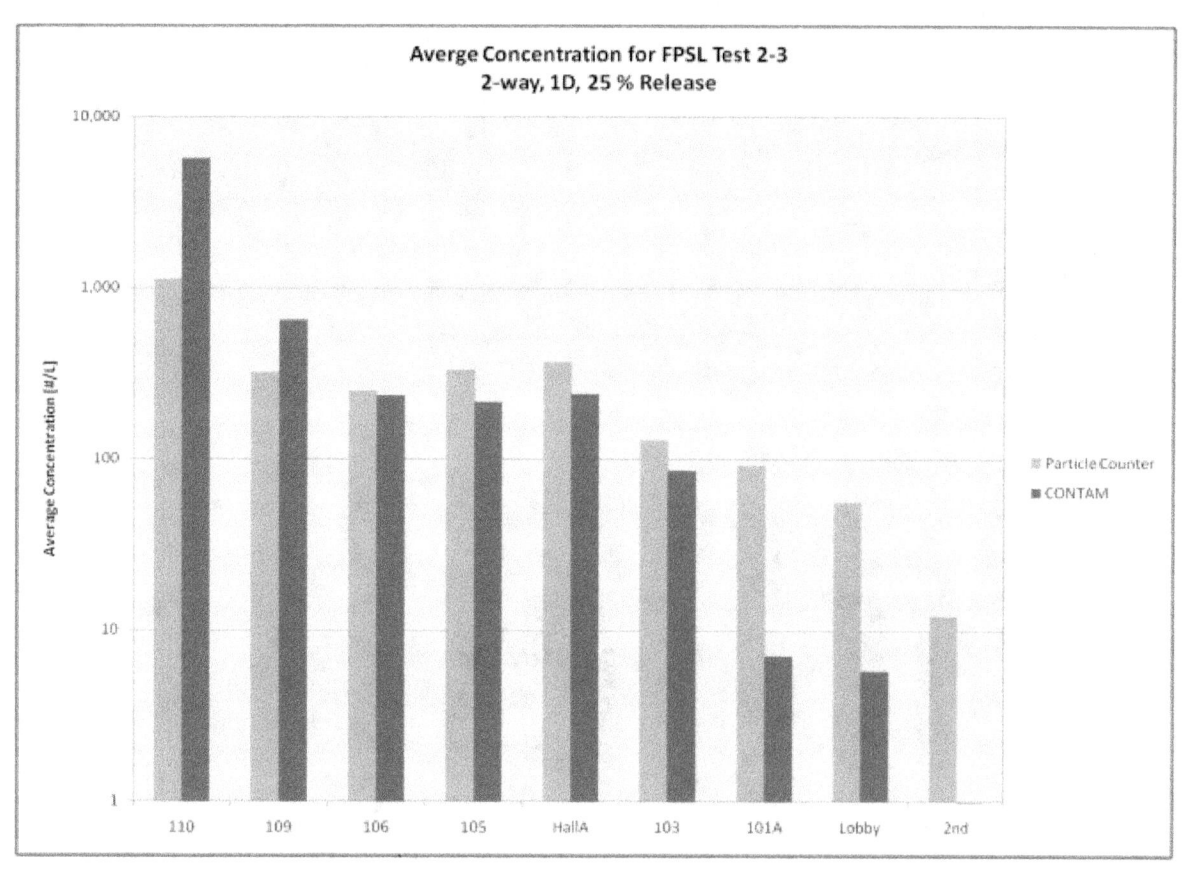

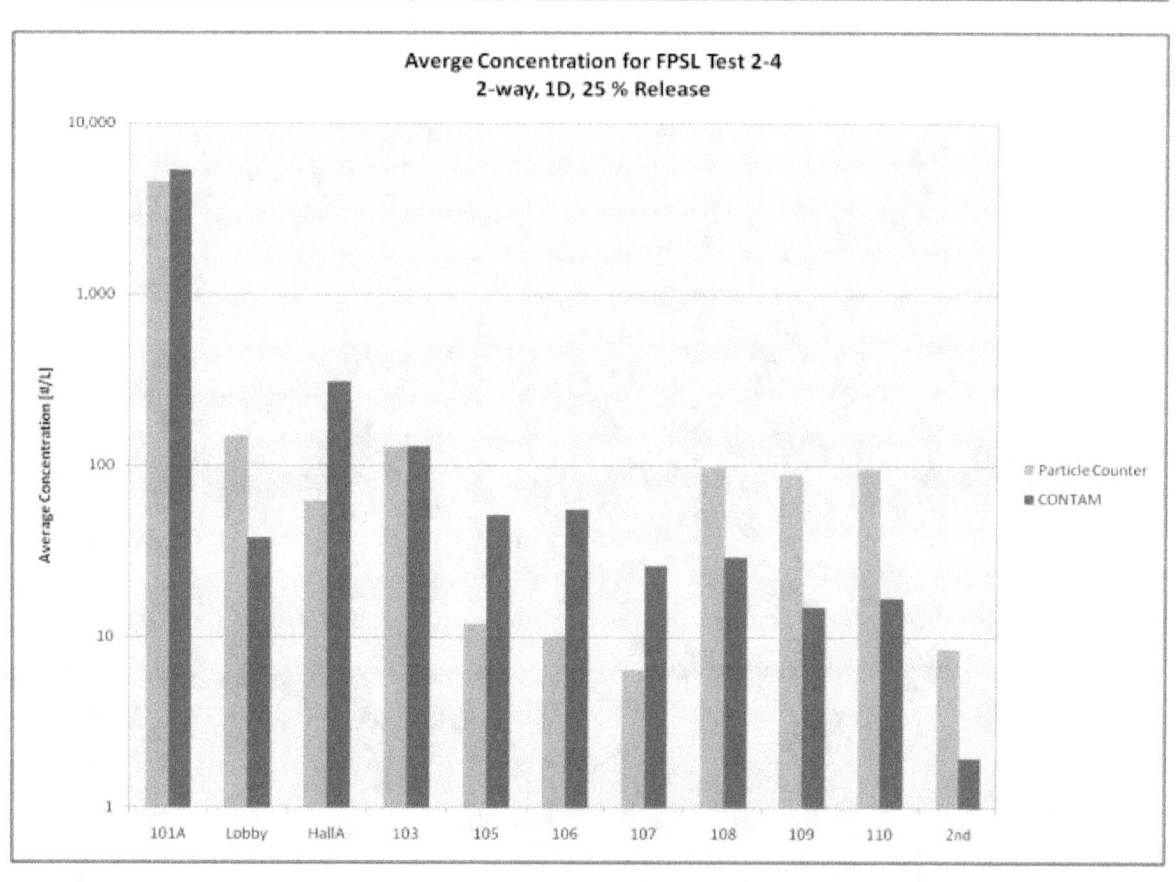

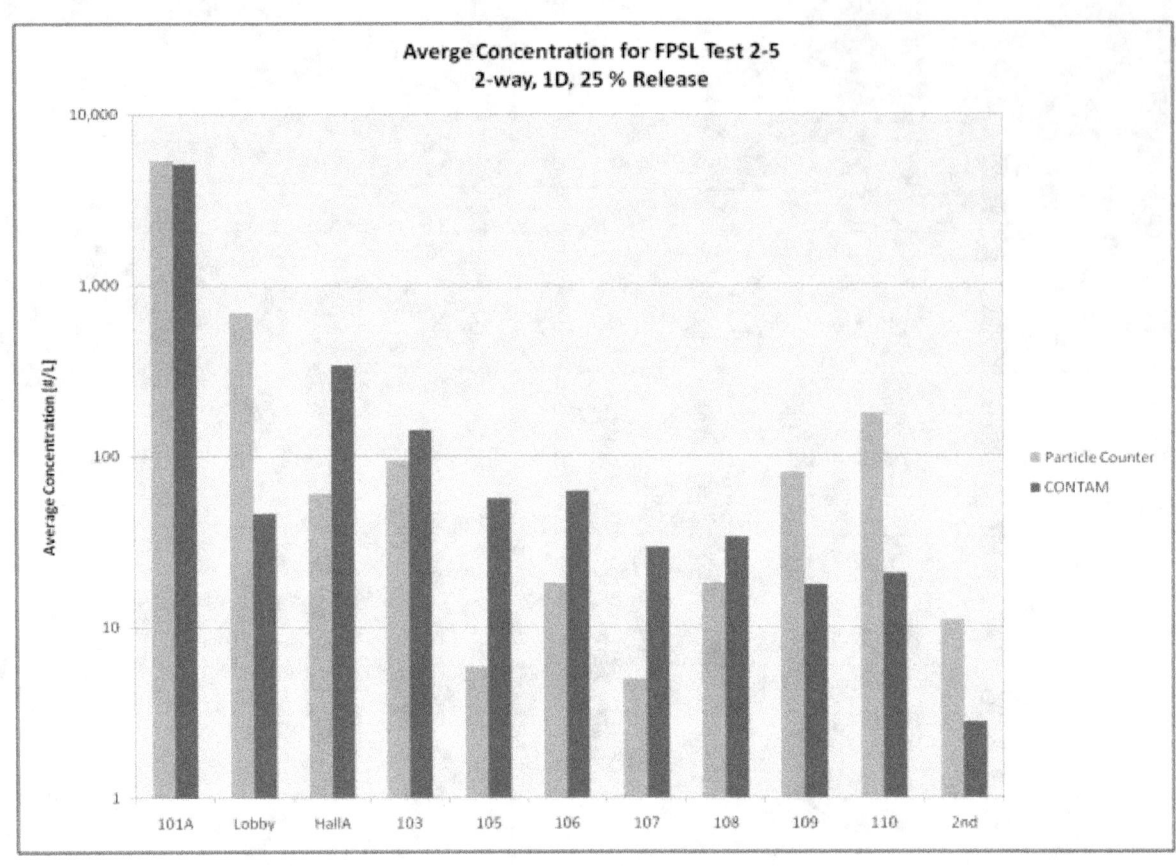

Average Concentration for FPSL Test 2-5
2-way, 1D, 25 % Release

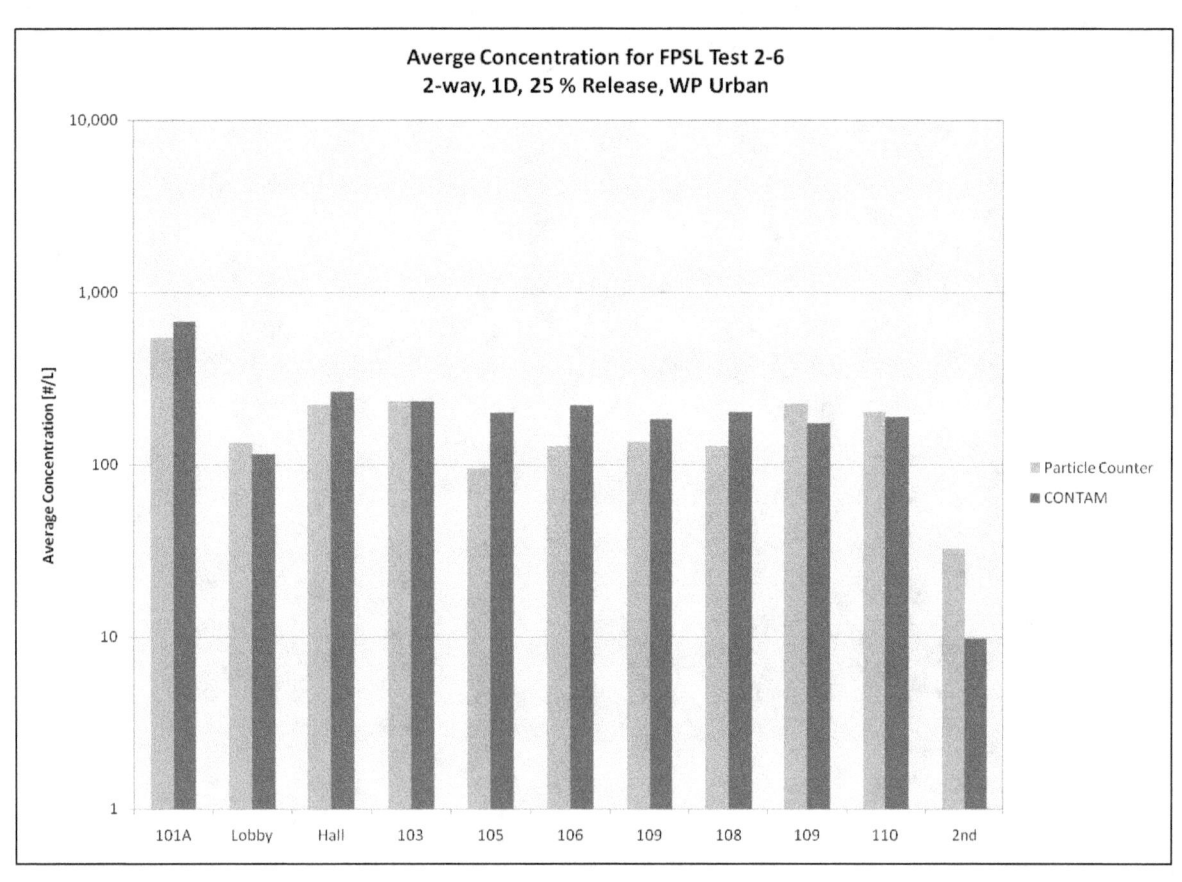

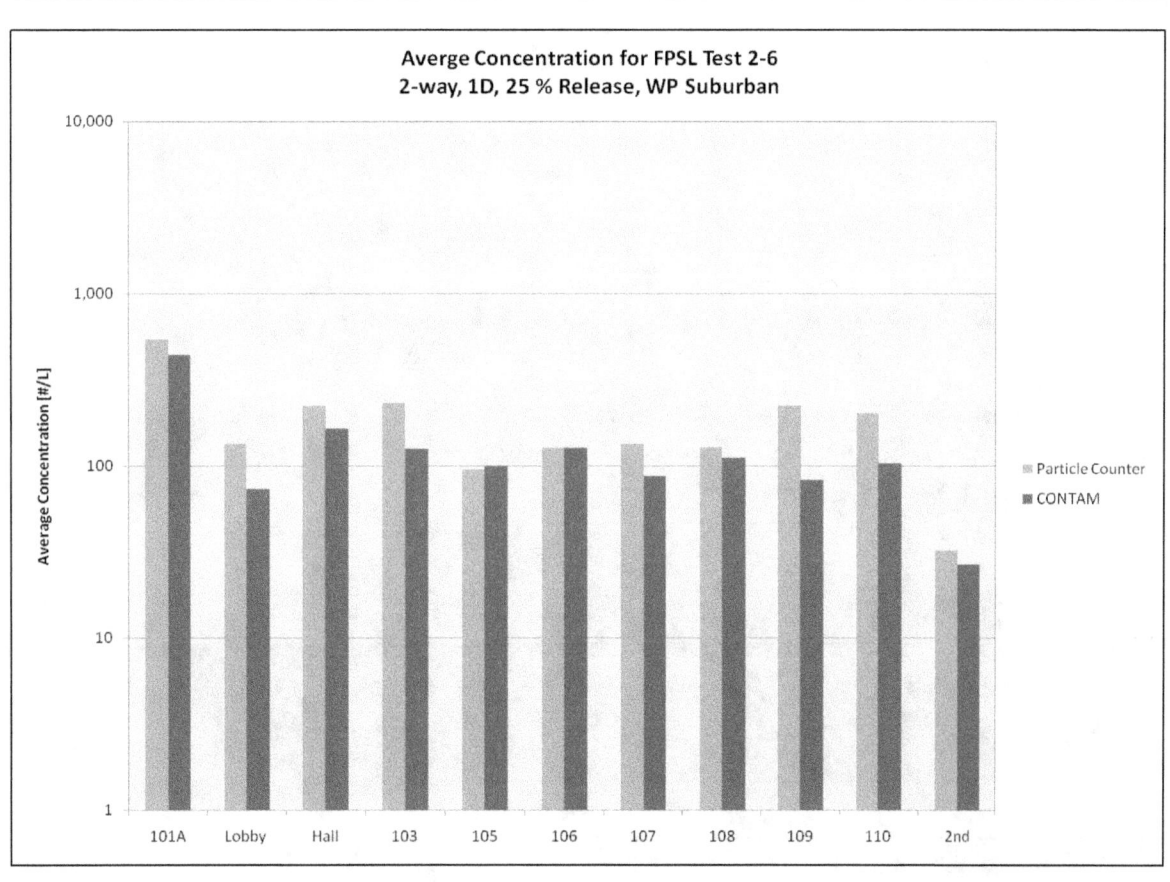

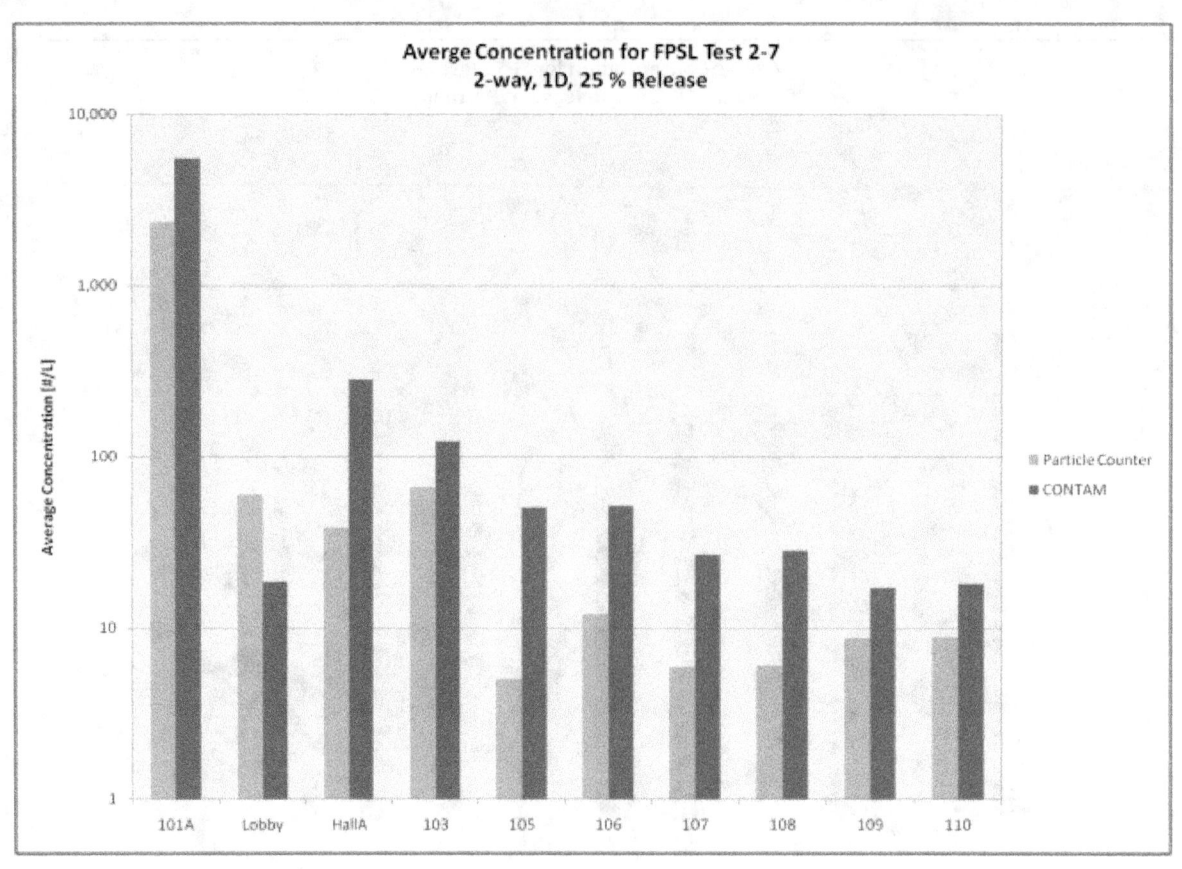

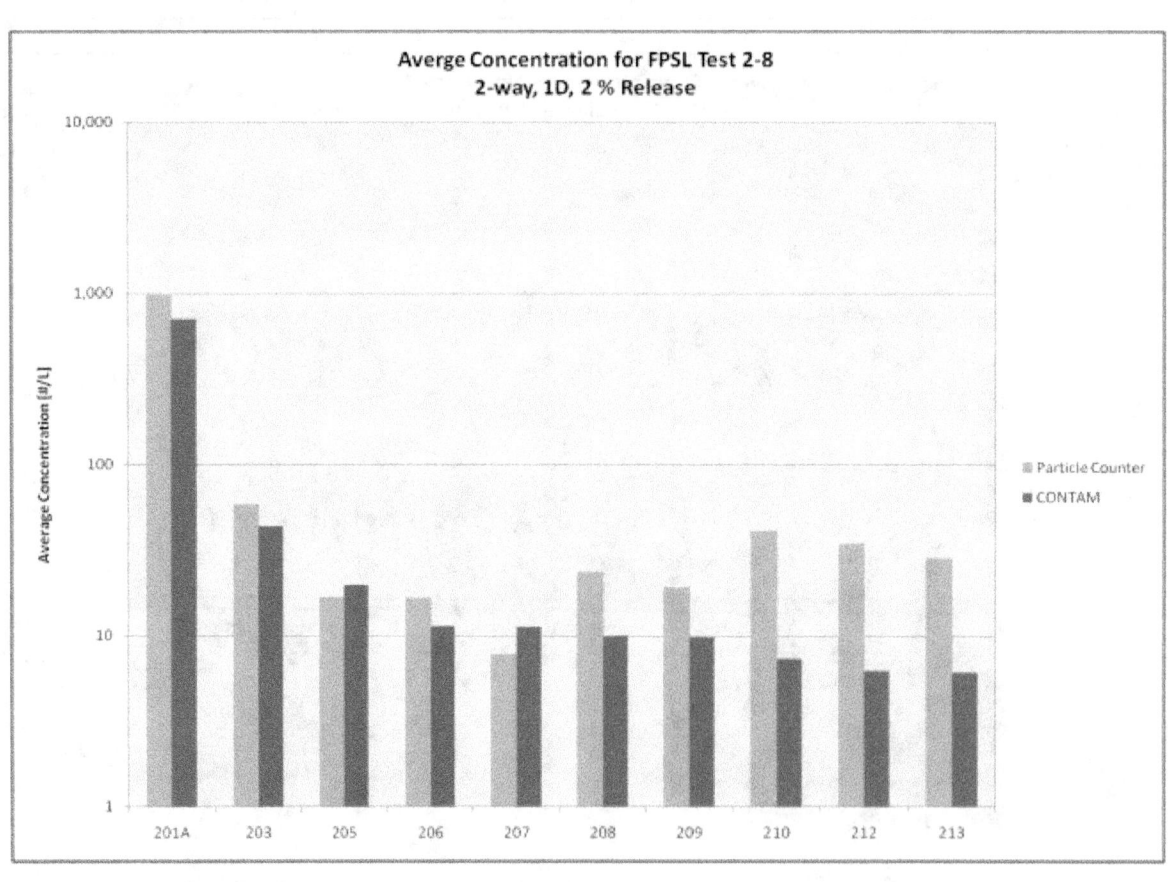

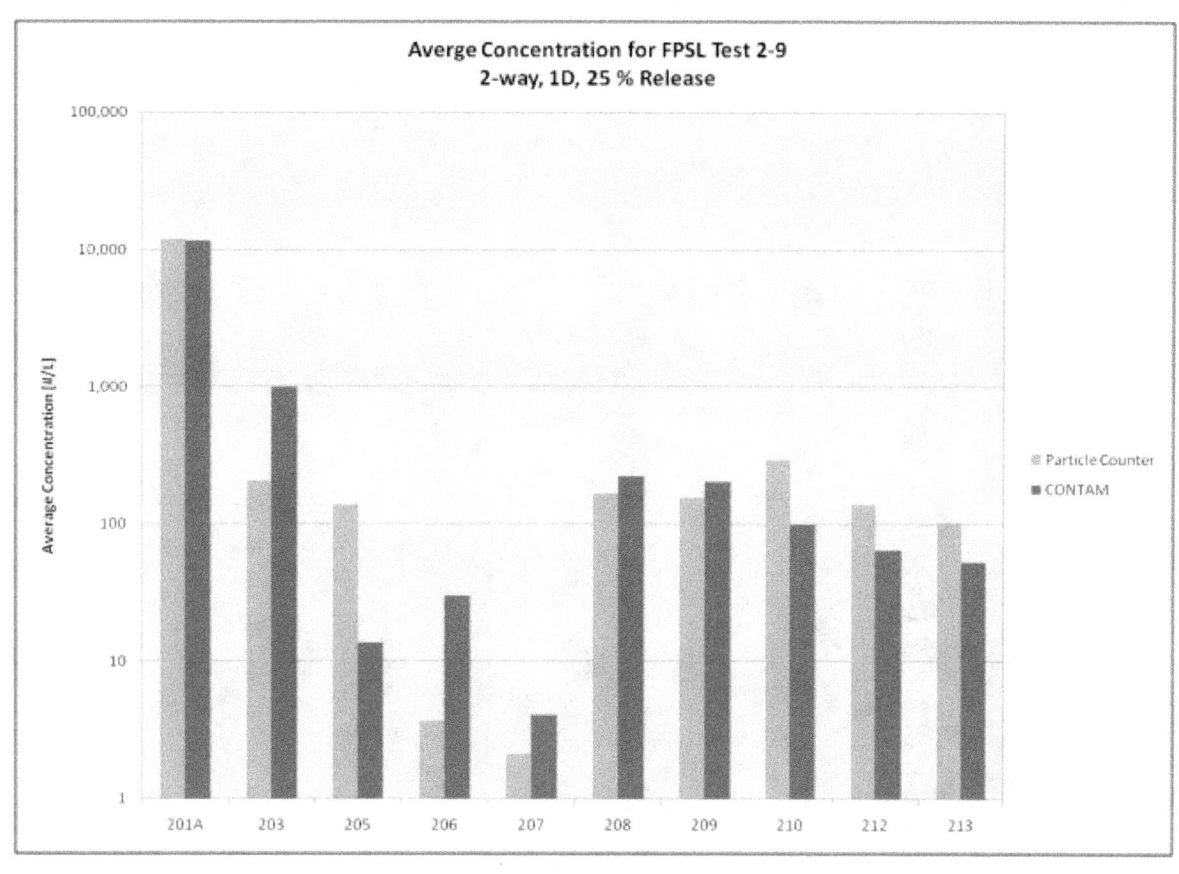

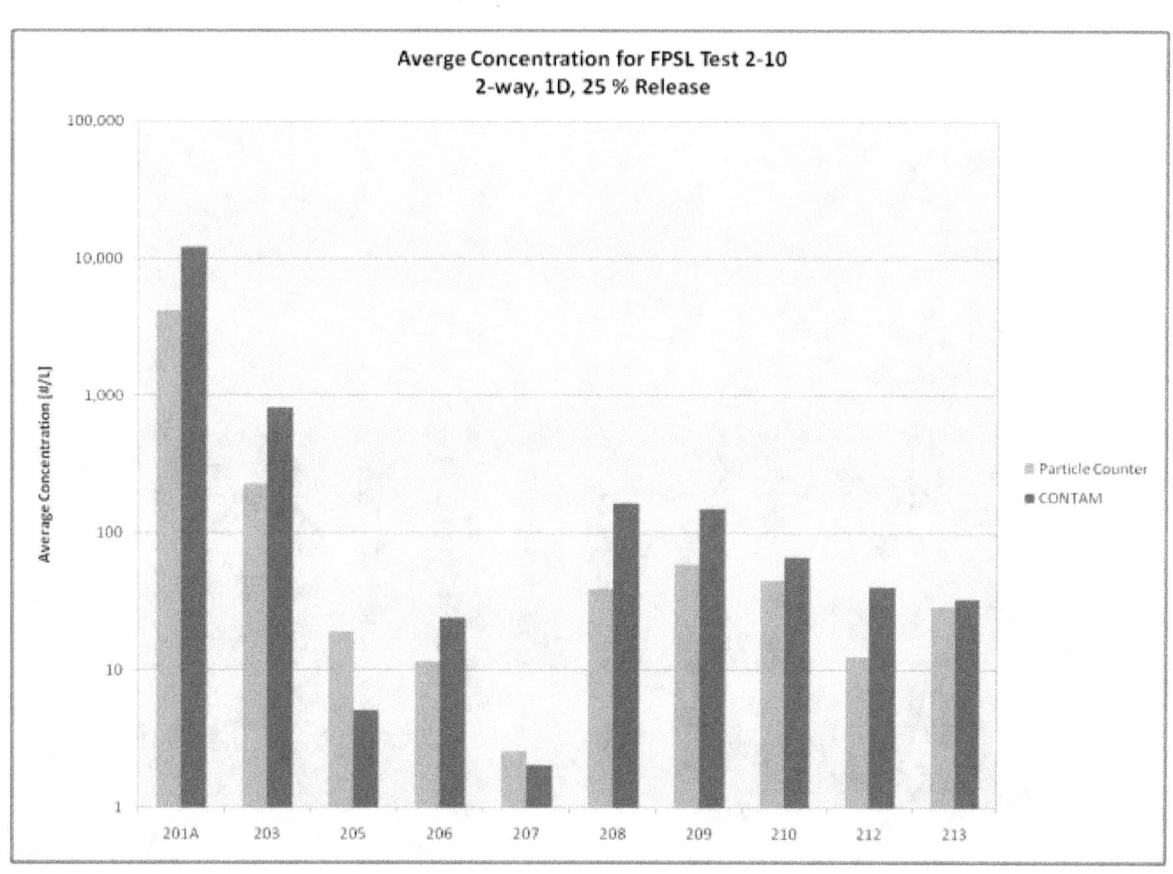

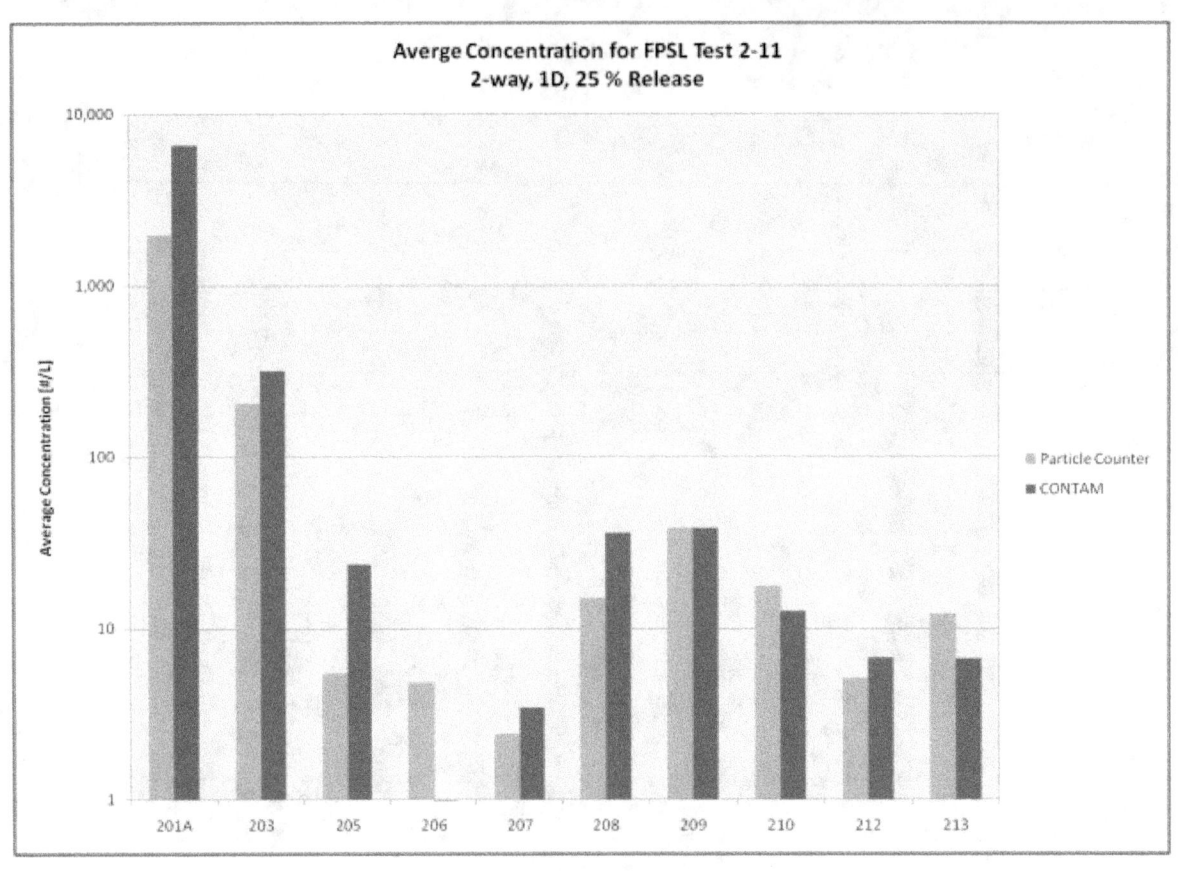

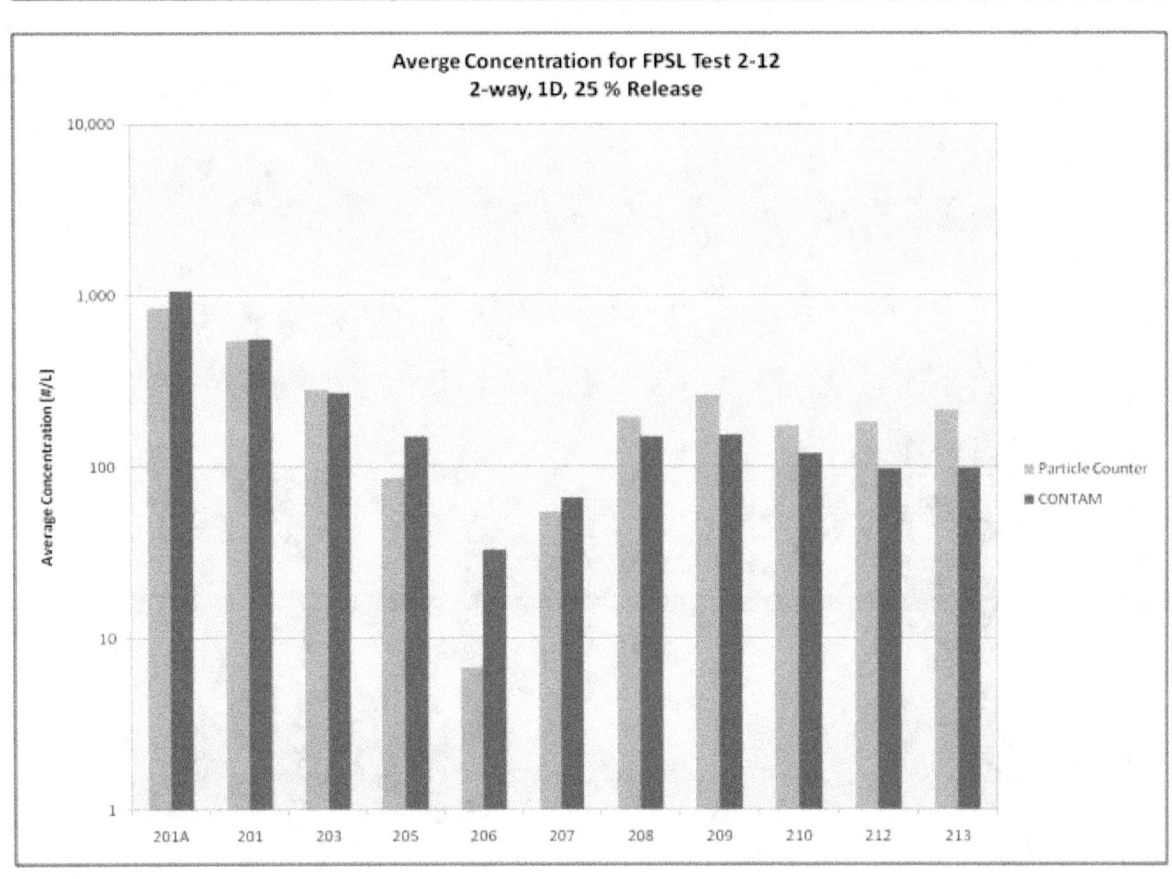

Appendix C – ASTM D5157 Results

This appendix provides tables displaying ASTM guide parameters of all building characterization tests. Simulations were performed employing the 1D convection-diffusion, two-way openings and reduced release amounts. The Mean Fractional Bias $|FB|$ provides a comparison between the average measured and predicted concentrations which can be directly related to the deposition within each zone assuming constant deposition rates and well-mixed zones. The first column of each table contains the room in which the release was located for the test, with columns arranged in order of increasing distance of physical proximity to the source zone, i.e., columns farthest to the right are farther away from the source zone.

INL-1 Characterization Tests

	Criteria	Lobby	Hall	101A	104	105	107	108	109	110
N		871	871	870	870	871	871	871	871	871
r	≥ 0.9	0.86	0.78	0.65	0.21	**0.93**	0.76	0.75	0.49	0.28
m	0.75 - 1.25	3.24	0.44	**1.04**	0.03	0.33	0.20	0.22	0.08	0.03
$b/\overline{Co}$	≤ 0.25	0.85	**0.14**	**0.14**	**0.15**	**0.01**	**0.03**	**0.06**	**0.07**	**0.08**
$NMSE$	≤ 0.25	4.86	0.54	0.31	6.12	1.63	3.29	2.45	6.28	10.26
$|FB|$	≤ 0.25	1.22	-0.53	**0.17**	-1.39	-0.99	-1.25	-1.12	-1.45	-1.60
$|FS|$	≤ 0.50	1.74	-1.01	0.87	-1.93	-1.56	-1.75	-1.67	-1.88	-1.95

Table 18 – Test 1-1

	Criteria	Lobby	Hall	101A	104	105	107	108	109	110
N		259	259	259	259	259	259	259	259	259
r	≥ 0.9	0.52	**0.90**	0.81	**0.95**	0.66	**0.90**	**0.92**	**0.97**	**0.92**
m	0.75 - 1.25	0.60	0.20	3.45	0.13	0.60	0.33	**0.97**	0.13	0.40
$b/\overline{Co}$	≤ 0.25	2.04	**0.20**	6.39	**0.03**	2.07	0.27	0.73	**0.03**	**0.25**
$NMSE$	≤ 0.25	1.14	1.90	10.21	7.53	2.20	1.77	0.46	9.95	1.18
$|FB|$	≤ 0.25	0.90	-0.86	1.63	-1.44	0.91	-0.51	0.52	-1.46	-0.42
$|FS|$	≤ 0.50	**0.31**	-1.81	1.79	-1.93	**-0.19**	-1.54	**0.10**	-1.93	-1.36

Table 19 – Test 1-2

	Criteria	Lobby	Hall	101A	104	105	107	108	109	110
N		193	193	193	193	193	192	193	193	193
r	≥ 0.9	-0.50	0.81	0.80	**0.94**	0.61	0.87	0.60	**0.96**	**0.91**
m	0.75 - 1.25	-0.57	0.32	16.40	0.50	23.31	0.30	0.45	0.08	0.16
$b/\overline{Co}$	≤ 0.25	3.53	0.63	26.68	**0.23**	1.76	0.31	1.28	**0.04**	**0.11**
$NMSE$	≤ 0.25	1.60	0.69	55.91	0.56	37.94	2.23	1.79	18.79	6.87
$|FB|$	≤ 0.25	0.99	**-0.06**	1.91	-0.32	1.85	-0.50	0.54	-1.58	-1.16
$|FS|$	≤ 0.50	**0.29**	-1.47	1.99	-1.13	2.00	-1.59	-0.56	-1.97	-1.88

Table 20 – Test 1-3

Criteria		Lobby	Hall	101A	104	105	107	108	109	110
N		4321	4321	4321	4321	4321	4321	4321	4321	4321
R	≥ 0.9	0.88	0.53	0.70	0.64	**0.92**	0.56	0.29	-0.49	-0.45
M	0.75 - 1.25	2.22	0.12	3.66	0.53	1.34	0.60	0.47	-0.29	-0.36
$b/\overline{Co}$	≤ 0.25	1.13	0.73	0.75	1.30	1.38	1.57	2.45	1.85	2.05
$NMSE$	≤ 0.25	6.09	2.77	5.31	0.60	1.16	0.82	1.56	1.14	0.96
$|FB|$	≤ 0.25	1.08	**-0.16**	1.26	0.59	0.92	0.74	0.98	0.44	0.52
$|FS|$	≤ 0.50	1.46	-1.80	1.86	**-0.38**	0.72	**0.14**	0.87	-1.00	**-0.43**

Table 21 – Test 1-4

Criteria		Lobby	Hall	101	101A	104	105	107	108	109	110
N		241	241	241	241	241	241	241	241	240	241
r	≥ 0.9	0.76	**0.98**	0.64	**0.97**	**0.95**	**0.97**	**0.99**	**0.94**	**0.98**	**0.99**
m	0.75 - 1.25	**0.99**	0.32	2.15	1.47	0.12	0.74	0.53	2.31	0.32	0.14
$b/\overline{Co}$	≤ 0.25	0.85	**0.02**	6.25	0.61	**0.01**	0.27	**0.10**	0.89	**0.09**	**0.00**
$NMSE$	≤ 0.25	0.46	2.04	7.77	0.75	11.27	**0.17**	0.74	2.91	2.91	11.75
$|FB|$	≤ 0.25	0.59	-0.99	1.57	0.70	-1.51	**0.01**	-0.45	1.05	-0.83	-1.50
$|FS|$	≤ 0.50	0.51	-1.62	1.68	0.79	-1.93	-0.52	-1.10	1.43	-1.61	-1.93

Table 22 – Test 1-5

Criteria		Lobby	Hall	101	101A	104	105	107	108	109	110
N		241	241	241	241	241	241	241	241	241	241
r	≥ 0.9	0.69	0.34	0.89	**0.98**	**0.99**	**0.98**	**0.97**	0.86	-0.08	0.32
m	0.75 - 1.25	**1.18**	0.07	1.64	**0.82**	0.48	3.07	9.66	27.57	-0.12	1.35
$b/\overline{Co}$	≤ 0.25	1.37	0.05	0.93	-0.08	0.11	-0.12	-2.52	-10.93	0.34	-0.89
$NMSE$	≤ 0.25	1.06	7.77	1.11	**0.12**	0.94	2.67	11.86	32.57	3.22	1.29
$|FB|$	≤ 0.25	0.87	-1.58	0.88	-0.29	-0.51	0.99	1.51	1.77	-1.28	-0.74
$|FS|$	≤ 0.50	0.99	-1.83	1.08	**-0.34**	-1.23	1.63	1.96	2.00	0.70	1.79

Table 23 – Test 1-6

Criteria		Lobby	Hall	101	101A	104	105	107	108	109	110
N		361	361	361	361	361	361	361	361	361	361
r	≥ 0.9	0.82	-0.26	0.41	0.85	**0.95**	**0.96**	**0.97**	**0.95**	**0.91**	0.80
m	0.75 - 1.25	**0.78**	-0.08	**0.90**	2.39	0.59	1.37	**0.90**	1.45	0.13	3.67
$b/\overline{Co}$	≤ 0.25	1.11	0.49	3.59	-0.49	0.05	0.13	0.06	0.72	0.57	-2.28
$NMSE$	≤ 0.25	0.46	1.81	2.87	0.72	0.37	0.28	**0.04**	0.85	0.32	0.36
$|FB|$	≤ 0.25	0.62	-0.84	1.27	0.62	-0.44	0.40	**-0.04**	0.74	-0.36	0.32
$|FS|$	≤ 0.50	**-0.09**	-1.67	1.31	1.55	-0.90	0.69	**-0.16**	0.79	-1.92	1.82

Table 24 – Test 1-7

	Criteria	Lobby	Hall	101	101A	104	105	107	108	109	110
N		223	223	223	223	223	223	223	223	223	223
r	≥ 0.9	0.65	0.59	0.79	**0.99**	0.63	**0.99**	**0.97**	**0.96**	-0.88	0.04
m	0.75 - 1.25	0.21	0.10	1.44	**1.01**	0.09	0.25	0.15	0.33	-0.03	0.01
$b/\overline{Co}$	≤ 0.25	0.93	0.06	2.84	0.15	0.01	0.05	0.05	0.24	0.69	1.01
$NMSE$	≤ 0.25	**0.18**	5.17	2.64	**0.03**	9.91	2.80	6.35	1.38	1.44	**0.05**
$\|FB\|$	≤ 0.25	**0.13**	-1.44	1.24	**0.15**	-1.63	-1.06	-1.35	-0.54	-0.42	**0.02**
$\|FS\|$	≤ 0.50	-1.61	-1.88	1.08	**0.04**	-1.92	-1.76	-1.91	-1.58	-1.99	-1.86

Table 25 – Test 1-8

	Criteria	201	201A	202	203	206	210	211	212	213	Hall
N		217	217	217	217	217	217	217	217	217	217
r	≥ 0.9	-0.11	0.89	0.81	0.12	0.69	**0.97**	**0.96**	**0.96**	**0.99**	0.42
m	0.75 - 1.25	-1.90	2.94	6.46	1.96	30.35	**1.01**	**1.19**	1.25	2.75	0.55
$b/\overline{Co}$	≤ 0.25	11.87	2.72	6.29	82.68	75.28	**0.20**	0.29	0.52	0.58	3.30
$NMSE$	≤ 0.25	13.38	4.07	12.55	90.91	141.88	**0.08**	**0.23**	0.47	2.52	2.43
$\|FB\|$	≤ 0.25	1.64	1.40	1.71	1.95	1.96	**0.19**	0.39	0.56	1.08	1.17
$\|FS\|$	≤ 0.50	1.99	1.67	1.94	1.98	2.00	**0.07**	**0.42**	0.53	1.54	0.52

Table 26 – Test 1-9

	Criteria	201	201A	202	203	206	210	211	212	213	Hall
N		1800	1801	1801	1801	1801	1801	1801	1801	1801	1801
r	≥ 0.9	**0.91**	0.87	0.86	0.89	0.85	-0.24	0.65	-0.54	-0.49	0.19
m	0.75 - 1.25	**1.16**	**0.90**	2.16	**1.09**	0.37	-0.05	0.26	-0.12	-0.27	0.02
$b/\overline{Co}$	≤ 0.25	**0.00**	**0.17**	**-0.01**	**-0.04**	**0.18**	0.36	**0.21**	0.48	0.92	0.47
$NMSE$	≤ 0.25	0.95	0.35	0.89	**0.07**	0.46	2.19	0.76	2.03	0.61	3.92
$\|FB\|$	≤ 0.25	**0.14**	**0.06**	0.73	**0.05**	-0.58	-1.06	-0.73	-0.94	-0.43	-0.68
$\|FS\|$	≤ 0.50	**0.48**	**0.07**	1.45	**0.40**	-1.37	-1.81	-1.46	-1.81	-1.05	-1.96

Table 27 – Test 1-10

	Criteria	201	201A	202	203	206	210	211	212	213	Hall
N		181	181	181	181	181	181	181	181	181	181
r	≥ 0.9	-0.06	0.74	0.21	0.61	0.78	**0.99**	**0.99**	**0.98**	**0.99**	0.71
m	0.75 - 1.25	-0.22	**1.00**	0.39	**0.79**	1.52	**0.92**	**1.16**	0.62	2.76	1.47
$b/\overline{Co}$	≤ 0.25	2.16	1.99	1.35	**0.18**	-0.73	**-0.09**	**-0.08**	**-0.01**	**0.11**	1.28
$NMSE$	≤ 0.25	1.43	1.44	0.71	**0.08**	**0.14**	**0.05**	**0.03**	0.40	2.01	1.29
$\|FB\|$	≤ 0.25	0.64	1.00	0.54	**-0.03**	**-0.24**	**-0.19**	**0.08**	-0.48	0.97	0.93
$\|FS\|$	≤ 0.50	1.75	0.59	1.07	0.52	1.16	**-0.15**	**0.32**	-0.86	1.55	1.24

Table 28 – Test 1-11

Criteria		201	201A	202	203	206	210	211	212	213	Hall
N		229	229	229	229	229	229	229	229	229	229
r	≥ 0.9	0.74	0.67	0.32	-0.05	0.86	**0.97**	**0.98**	0.86	**0.99**	-0.27
m	0.75 - 1.25	2.00	**1.03**	0.52	-0.07	5.66	1.34	1.68	1.86	4.63	-0.23
$b/\overline{Co}$	≤ 0.25	**0.16**	1.05	0.77	1.88	1.86	**-0.05**	**-0.17**	-0.27	**0.15**	1.86
$NMSE$	≤ 0.25	1.30	0.65	0.28	0.63	7.94	**0.15**	0.33	0.60	4.84	0.71
$|FB|$	≤ 0.25	0.74	0.70	0.26	0.58	1.53	0.26	0.41	0.45	1.31	0.48
$|FS|$	≤ 0.50	1.51	0.81	0.89	0.72	1.91	0.62	0.99	1.29	1.82	**-0.33**

Table 29 – Test 1-12

Criteria		201	201A	202	203	206	210	211	212	213	Hall
N		217	217	217	217	217	217	217	217	217	217
r	≥ 0.9	0.75	0.47	0.81	**0.93**	0.81	**0.94**	**0.99**	0.88	**0.98**	0.58
m	0.75 - 1.25	1.59	0.60	4.19	1.55	9.35	2.93	1.45	6.79	2.79	**1.13**
$b/\overline{Co}$	≤ 0.25	0.34	1.03	5.08	0.85	1.43	1.28	**-0.02**	1.41	-0.18	1.55
$NMSE$	≤ 0.25	0.93	0.34	8.63	0.89	13.10	3.61	**0.20**	10.08	1.70	1.25
$|FB|$	≤ 0.25	0.63	0.48	1.61	0.82	1.66	1.23	0.35	1.57	0.89	0.92
$|FS|$	≤ 0.50	1.27	**0.49**	1.86	0.95	1.97	1.63	0.72	1.93	1.56	1.17

Table 30 – Test 1-13

INL-2 Characterization Tests

Criteria		101A	Lobby	Hall	103	105	106	109	110	2nd
N		301	301	301	301	301	301	301	301	301
r	≥ 0.9	**0.97**	0.89	0.74	**0.93**	**0.99**	0.89	**0.98**	0.89	0.48
m	0.75 - 1.25	1.50	0.10	**0.95**	**0.93**	1.72	0.72	0.46	0.43	0.25
$b/\overline{Co}$	≤ 0.25	**0.27**	**0.00**	1.03	0.45	0.32	**0.10**	-0.07	-0.10	**0.17**
$NMSE$	≤ 0.25	0.38	19.33	0.60	**0.16**	0.70	**0.16**	1.26	1.64	2.03
$|FB|$	≤ 0.25	0.55	-1.63	0.66	0.32	0.68	**-0.19**	-0.87	-1.00	-0.83
$|FS|$	≤ 0.50	0.83	-1.95	**0.49**	**-0.01**	1.01	**-0.42**	-1.28	-1.25	-1.16

Table 31 – Test 2-1

Criteria		101A	Lobby	Hall	103	105	106	109	110	2nd
N		241	241	241	241	241	241	241	241	241
r	≥ 0.9	0.40	**0.90**	0.45	**0.96**	0.29	**0.94**	**0.91**	0.87	0.19
m	0.75 - 1.25	0.58	0.53	0.62	**1.23**	0.32	**0.96**	0.36	0.53	0.37
$b/\overline{Co}$	≤ 0.25	4.56	**0.12**	2.49	**0.16**	0.77	0.33	**-0.12**	**-0.11**	0.30
$NMSE$	≤ 0.25	3.74	0.39	1.54	**0.15**	0.53	**0.12**	2.79	1.01	0.75
$|FB|$	≤ 0.25	1.35	-0.41	1.03	0.32	**0.08**	0.25	-1.22	-0.82	-0.40
$|FS|$	≤ 0.50	0.70	-0.95	0.64	**0.49**	0.25	0.04	-1.46	-0.91	1.18

Table 32 – Test 2-2

	Criteria	110	109	106	105	Hall	103	101A	Lobby	2nd
N		181	181	181	181	181	181	181	181	181
r	≥ 0.9	0.53	0.86	0.72	**0.92**	0.85	0.87	0.53	0.78	-0.28
m	0.75 - 1.25	**0.89**	**0.93**	0.35	0.61	0.30	0.64	0.19	0.16	-0.03
$b/\overline{Co}$	≤ 0.25	4.17	1.09	0.58	**0.04**	0.35	**0.02**	**-0.11**	**-0.06**	**0.05**
$NMSE$	≤ 0.25	3.69	0.65	0.82	0.32	0.85	0.37	11.46	9.49	42.77
$\|FB\|$	≤ 0.25	1.34	0.68	**-0.06**	-0.42	-0.42	-0.41	-1.71	-1.62	-1.90
$\|FS\|$	≤ 0.50	0.96	**0.15**	-1.22	-0.79	-1.55	-0.61	-1.56	-1.84	-1.96

Table 33 – Test 2-3

	Criteria	101A	Lobby	Hall	103	105	106	109	110	2nd
N		241	241	241	241	241	241	241	241	241
r	≥ 0.9	0.65	**0.91**	0.88	**0.92**	**0.95**	0.74	0.67	0.84	-0.03
m	0.75 - 1.25	0.20	0.26	2.86	**0.96**	3.54	4.65	0.14	0.19	-0.02
$b/\overline{Co}$	≤ 0.25	0.98	**0.00**	2.09	**0.06**	0.75	0.95	**0.02**	**-0.01**	**0.25**
$NMSE$	≤ 0.25	0.64	3.61	3.98	**0.10**	4.46	7.31	6.93	6.22	3.56
$\|FB\|$	≤ 0.25	**0.17**	-1.19	1.33	**0.01**	1.24	1.39	-1.42	-1.40	-1.25
$\|FS\|$	≤ 0.50	-1.67	-1.70	1.65	**0.08**	1.73	1.90	-1.82	-1.81	-0.83

Table 34 – Test 2-4

	Criteria	101A	Lobby	Hall	103	105	106	109	110	2nd
N		241	241	241	241	241	241	241	241	241
r	≥ 0.9	0.69	0.10	0.82	**0.93**	0.68	0.17	0.70	0.66	-0.44
m	0.75 - 1.25	0.16	0.01	4.60	1.86	7.85	1.69	0.22	0.11	-0.28
$b/\overline{Co}$	≤ 0.25	0.80	**0.06**	1.05	-0.34	2.02	1.82	**0.00**	**0.00**	0.54
$NMSE$	≤ 0.25	1.00	22.58	4.94	0.47	14.57	4.33	4.12	9.90	3.62
$\|FB\|$	≤ 0.25	**-0.04**	-1.75	1.40	0.41	1.63	1.11	-1.27	-1.59	-1.19
$\|FS\|$	≤ 0.50	-1.80	-1.98	1.88	1.19	1.97	1.96	-1.63	-1.89	-0.82

Table 35 – Test 2-5

	Criteria	101A	Lobby	Hall	103	105	106	107	108	109	110	2nd
N		4321	4321	4321	4321	4321	4321	4321	4321	4321	4321	4321
r	≥ 0.9	0.74	**0.91**	0.88	**0.96**	**0.96**	**0.97**	**0.93**	**0.92**	0.50	0.74	0.69
m	0.75 - 1.25	0.61	0.38	0.60	0.71	**1.20**	**1.03**	**0.79**	**0.85**	0.19	0.44	0.44
$b/\overline{Co}$	≤ 0.25	0.63	0.49	0.59	0.30	0.90	0.68	0.56	0.72	0.58	0.50	1.02
$NMSE$	≤ 0.25	3.17	0.35	0.32	**0.08**	0.63	0.32	**0.14**	0.27	0.82	0.31	0.39
$\lvert FB \rvert$	≤ 0.25	**0.21**	-0.15	**0.17**	**0.00**	0.71	0.52	0.30	0.44	-0.26	**-0.06**	0.37
$\lvert FS \rvert$	≤ 0.50	**-0.39**	-1.41	-0.74	-0.60	**0.45**	**0.12**	**-0.33**	**-0.16**	-1.48	-0.97	-0.85

Table 36 – Test 2-6 (WP urban)

	Criteria	101A	Lobby	Hall	103	105	106	107	108	109	110	2nd
N		4321	4321	4321	4321	4321	4321	4321	4321	4321	4321	4321
r	≥ 0.9	0.72	0.84	0.83	0.79	0.84	0.67	**0.96**	0.78	0.88	0.81	0.08
m	0.75 - 1.25	0.54	0.65	0.59	**0.91**	**0.94**	0.49	**0.89**	0.26	0.52	0.44	0.07
$b/\overline{Co}$	≤ 0.25	0.27	**0.09**	**-0.05**	**0.14**	**0.05**	**0.16**	**-0.02**	**0.10**	**-0.01**	**0.10**	0.76
$NMSE$	≤ 0.25	4.95	0.60	0.75	0.29	0.27	0.60	**0.09**	2.30	0.84	0.89	1.41
$\lvert FB \rvert$	≤ 0.25	**-0.21**	-0.30	-0.59	**0.05**	**-0.01**	-0.43	**-0.14**	-0.92	-0.65	-0.59	**-0.19**
$\lvert FS \rvert$	≤ 0.50	-0.56	-0.51	-0.65	0.28	**0.24**	-0.62	**-0.14**	-1.59	-0.97	-1.08	**-0.22**

Table 37 – Test 2-6 (WP suburban)

	Criteria	101A	Lobby	Hall	103	105	106	107	108	109	110	2nd
N		241	241	241	241	241	241	240	241	241	241	241
r	≥ 0.9	0.33	**0.97**	0.80	0.89	-0.12	0.54	0.14	0.39	0.33	0.43	0.34
m	0.75 - 1.25	0.11	0.36	6.62	1.27	-2.19	4.88	**1.24**	3.14	1.63	2.02	0.06
$b/\overline{Co}$	≤ 0.25	2.26	**-0.05**	0.84	0.60	12.35	-0.59	3.34	1.56	0.35	**0.05**	**0.01**
$NMSE$	≤ 0.25	1.77	2.39	7.47	0.71	16.53	5.60	6.97	6.89	2.33	2.38	16.26
$\lvert FB \rvert$	≤ 0.25	0.81	-1.05	1.53	0.61	1.64	1.24	1.28	1.30	0.66	0.70	-1.76
$\lvert FS \rvert$	≤ 0.50	-1.60	-1.52	1.94	0.68	1.99	1.95	1.95	1.94	1.84	1.83	-1.88

Table 38 – Test 2-7

	Criteria	201A	203	205	206	207	208	209	210	212	213
N		241	241	241	241	241	241	241	241	241	241
r	≥ 0.9	0.44	0.46	0.77	**0.92**	0.89	**0.90**	**0.94**	0.76	0.81	0.82
m	0.75 - 1.25	0.06	0.52	**0.85**	0.41	0.67	0.22	0.24	0.05	0.03	0.03
$b/\overline{Co}$	≤ 0.25	0.66	**0.23**	0.33	0.27	0.78	**0.20**	0.27	**0.13**	**0.15**	**0.19**
$NMSE$	≤ 0.25	3.07	0.27	**0.17**	0.37	**0.23**	1.32	0.99	5.09	6.04	5.23
$\lvert FB \rvert$	≤ 0.25	-0.32	-0.29	**0.17**	-0.38	0.37	-0.81	-0.65	-1.39	-1.39	-1.29
$\lvert FS \rvert$	≤ 0.50	-1.92	**0.25**	**0.18**	-1.33	-0.55	-1.77	-1.75	-1.98	-1.99	-2.00

Table 39 – Test 2-8

	Criteria	201A	203	205	206	207	208	209	210	212	213
N		361	361	361	361	361	361	361	361	361	361
r	≥ 0.9	0.83	0.84	**0.91**	-0.09	0.17	**0.95**	**0.98**	**0.93**	**0.96**	**0.98**
m	0.75 - 1.25	0.34	4.62	0.08	-1.31	0.30	1.49	1.35	0.43	0.51	0.44
$b/\overline{Co}$	≤ 0.25	0.64	**0.21**	**0.02**	9.43	1.62	**-0.13**	**-0.02**	**-0.09**	**-0.05**	**0.07**
$NMSE$	≤ 0.25	0.51	3.87	13.31	15.90	1.47	**0.24**	**0.16**	1.69	0.93	1.07
$|FB|$	≤ 0.25	**-0.02**	1.31	-1.64	1.56	0.63	0.30	0.28	-0.98	-0.73	-0.64
$|FS|$	≤ 0.50	-1.43	1.87	-1.97	1.98	0.99	0.83	0.61	-1.30	-1.12	-1.33

Table 40 – Test 2-9

	Criteria	201A	203	205	206	207	208	209	210	212	213
N		301	301	301	301	301	301	301	301	301	301
r	≥ 0.9	0.69	0.56	-0.41	-0.17	0.08	**0.96**	**0.98**	**0.90**	0.63	**0.96**
m	0.75 - 1.25	0.55	3.43	-0.43	-0.78	0.19	4.73	2.50	2.78	3.65	**1.20**
$b/\overline{Co}$	≤ 0.25	2.38	**0.20**	0.70	2.87	0.61	-0.52	**0.06**	-1.29	-0.38	**-0.06**
$NMSE$	≤ 0.25	1.64	2.82	4.89	2.29	3.57	4.21	1.63	0.77	3.56	**0.12**
$|FB|$	≤ 0.25	0.98	1.13	-1.15	0.71	**-0.22**	1.23	0.88	0.40	1.06	**0.13**
$|FS|$	≤ 0.50	**-0.45**	1.90	**0.12**	1.81	1.41	1.84	1.47	1.62	1.89	**0.45**

Table 41 – Test 2-10

	Criteria	201A	203	205	206	207	208	209	210	212	213
N		241	241	241	241	241	241	241	241	241	241
r	≥ 0.9	0.69	0.27	-0.17	-0.10	0.48	**0.92**	**0.98**	0.84	0.41	**0.94**
m	0.75 - 1.25	1.59	0.61	-1.60	-0.01	**0.84**	2.04	**1.02**	**0.83**	**0.86**	0.51
$b/\overline{Co}$	≤ 0.25	1.75	0.95	6.01	**0.10**	0.59	0.38	**-0.01**	**-0.12**	0.47	**0.04**
$NMSE$	≤ 0.25	1.88	0.70	9.29	13.99	1.33	1.45	**0.03**	0.30	1.01	0.85
$|FB|$	≤ 0.25	1.08	0.44	1.26	-1.69	0.36	0.83	**0.01**	-0.33	0.29	-0.58
$|FS|$	≤ 0.50	1.36	1.37	1.95	-1.95	1.04	1.33	**0.08**	**-0.02**	1.27	-1.09

Table 42 – Test 2-11

	Criteria	201A	201	203	205	206	207	208	209	210	212	213
N		4591	4591	4591	4591	4591	4591	4591	4591	4591	4591	4591
r	≥ 0.9	0.85	**0.92**	0.78	-0.02	0.54	0.79	**0.94**	**0.96**	0.59	0.83	0.88
m	0.75 - 1.25	**1.02**	1.50	**1.00**	-0.04	4.59	**0.92**	**0.83**	0.71	0.61	0.55	0.50
$b/\overline{Co}$	≤ 0.25	0.49	-0.28	**0.15**	2.14	1.20	0.53	**0.09**	**-0.01**	0.22	**0.09**	**0.05**
$NMSE$	≤ 0.25	2.61	0.87	0.41	1.14	7.35	**0.24**	**0.06**	**0.19**	**0.21**	0.31	0.49
$|FB|$	≤ 0.25	0.41	**0.20**	**0.14**	0.71	1.41	0.37	**-0.08**	-0.34	**-0.18**	-0.43	-0.58
$|FS|$	≤ 0.50	**0.36**	0.91	**0.49**	1.12	1.95	**0.32**	**-0.26**	-0.57	**0.07**	-0.78	-1.01

Table 43 – Test 2-12

Appendix D – INL Characterization Weather

This appendix provides plots of the weather conditions during the two sets of building characterization tests.

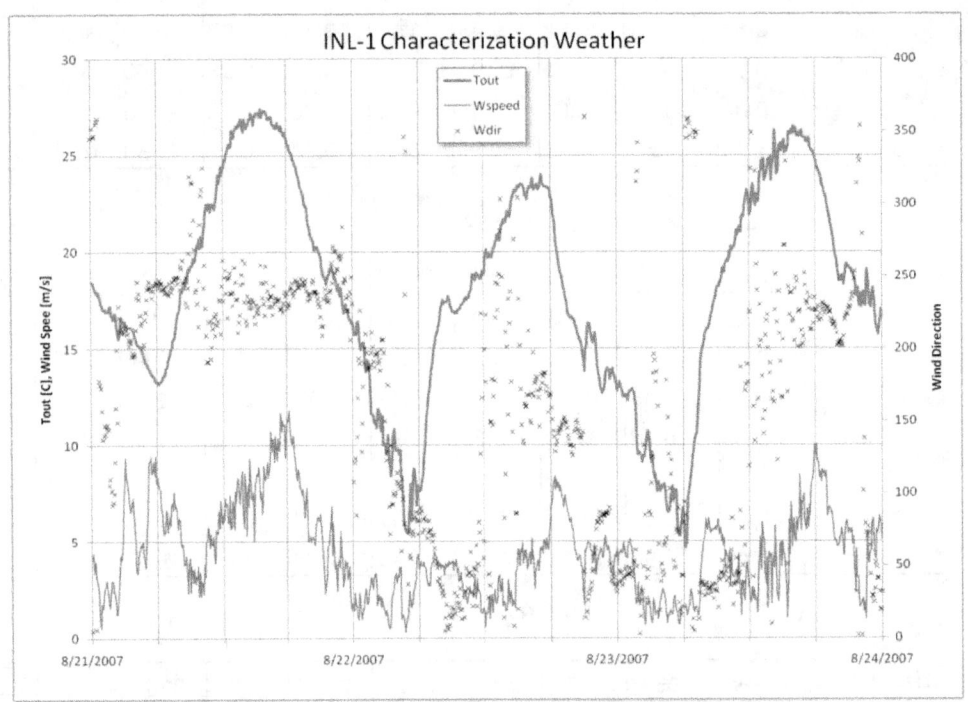

Figure 53 – INL-1 Characterization Weather

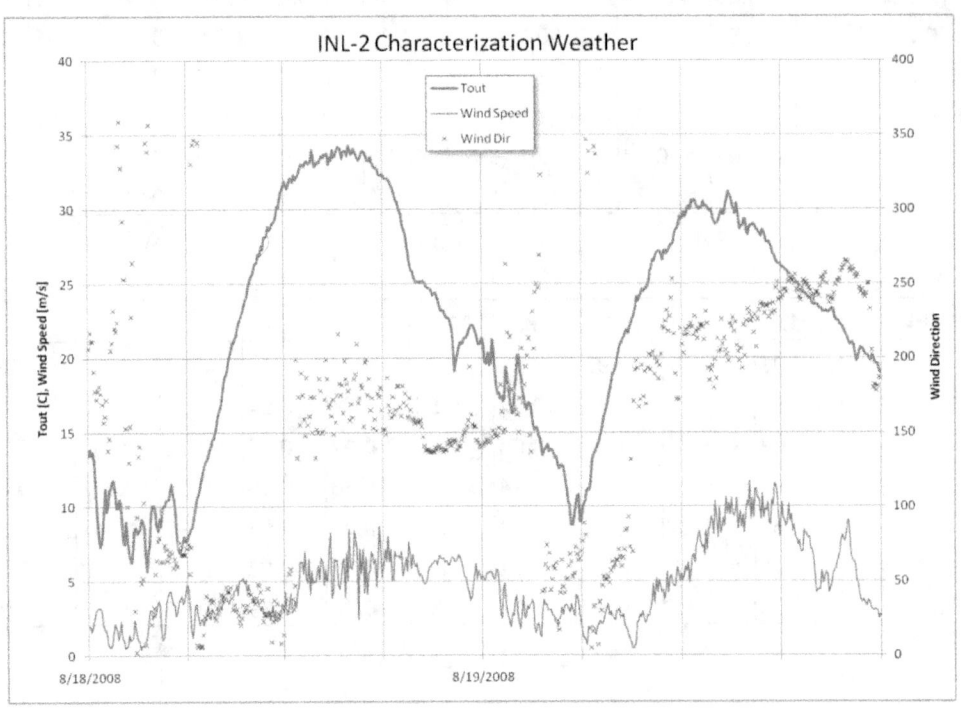

Figure 54 – INL-2 Characterization Weather

www.ingramcontent.com/pod-product-compliance
Lightning Source LLC
Chambersburg PA
CBHW081827170526
45167CB00007B/2737